Book WITHDRAWN

You will find basics about the safe handling of specific foods listed under the name of the food itself; for example, Chicken.

If the information inspires you to find out about *Salmonella*, you can then read about that pathogen in the Sources of Foodborne Illness section.

And if you want to find out about certain methods like casserole cooking, you can look them up under Casseroles.

There are only two big words you need to know—"microbiology" and "pathogenic."

We talk a lot about microbiology, but don't let it worry you. We need to talk about it because microbes cause the majority of food-related illness.

If you keep in mind that the word pathogenic means disease-producing or disease-causing, you'll be able to make sense of the food sciences found in this book.

The "Danger Zone," meanwhile, is the temperature range where pathogens will multiply rapidly if conditions are favorable.

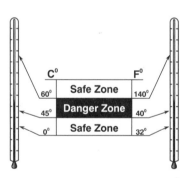

Tools you will need:

- ► probe thermometer
- ► spray bottle
- ► 1 clean cloth
- ► 1 gallon bucket for sanitizer
- ► household chlorine bleach

**Safe Food Rule Number One:
If in doubt, throw it out!**

GOOD EATING!

Everybody's FOODSAFE Kitchen
by Sheri Nielson
Edited by Wm (Bill) Hines C.P.H.I. (c) & Gerry Penner C.P.H.I. (c)
Illustrations by Bill W. Elford
Cover & Page Design - Latent Image Communications

Canadian Cataloguing in Publication Data

Nielson, Sheri, 1954-
    Everybody's foodsafe kitchen

Includes bibliographical references and index.
ISBN 0-9681588-0-3

    1. Food handling. 2. Food contamination--Prevention.
3. Food--Microbiology. I. Hines, Bill. II. Penner, Gerry.
III. Title.
TX601.N53 1996   641.3   C96-910808-7

Published by:
Everybody's Kitchen Ventures Ltd.
Salt Spring Island, British Columbia, Canada.

 Printed and Distributed by:
The Open Learning Agency
Burnaby, British Columbia, Canada

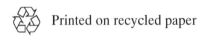 Printed on recycled paper

Everybody's SAFE Kitchen

Your Step-by-Step Guide to the Safe Preparation of Food

By Sheri Nielson

Edited by Wm (Bill) Hines C.P.H.I. (c) & Gerry Penner C.P.H.I. (c)

Illustrations by Bill W. Elford

EKV

Acknowledgements

Thank you to all of the people who worked so hard on this project.

Thank you to the many people in government, educational and food industry agencies in Canada and the United States who patiently fielded questions and offered their expertise and suggestions.

Thank you to Dr. Brent Skura at the Department of Food Science, Faculty of Agricultural Sciences, University of British Columbia who always found time to clarify and verify information, direct referrals and educate the author who learned once again... the more you learn, the less you know.

Thank you to Dr. Ewen Todd and his team at the Bureau of Microbial Hazards, Health Protection Branch, Health Canada for providing answers when we could not find a solid source.

A special acknowledgement in memory of Barry Black. Testimony to his contribution as project co-ordinator and educator lives on in the food industry people who learn about food safety in British Columbia, Canada and World-Wide through the FOODSAFE Course.

Thank you to the FOODSAFE Steering Committee for their on going support and encouragement.

To my editors, who have acted as resource people through my career as a FOODSAFE instructor and now as an author of a non-technical book on a very technical subject, thank you.

CONTENTS

Preface

Introduction

Preface

We all have many questions when it comes to food safety. No one can better answer your questions than my friend and associate Sheri Nielson. In this book, Sheri looks at the reality of what we should be concerned with in our day-to-day kitchen activities.

Certainly we must be alarmed when pesticides and herbicides are improperly applied to our food chain crops; however, these substances are, for the most part, used with care and rank least on the list of acute problems affecting food integrity.

Investigation has proven that microbes and people are the prime causes of foodborne illness.

The tiniest forms of life on our planet cause us the greatest food protection grief. Their adaptability and quick regeneration times—when conditions are ideal—allow these microscopic pathogens to adjust to our protective measures with alarming frequency.

By following the principles recommended in this book you will be armed with the defenses necessary to protect your families against foodborne illness.

Prevention in the first instance must be the rule. Treatment of foodborne illness can often be ineffective, or indeed, nonexistent without a quick and accurate diagnosis.

These illnesses could be considered as the "Great Pretenders" to the medical community.

Microorganisms have been a part of our environment since before homo sapiens inhabited the earth. Microbes are able to adapt quite readily to subtle change. For example, some can and have become resistant to many antibiotics.

Sheri Nielson has become a specialist in food sanitation training and other disciplines associated with the food service and hospitality industry. Her natural curiosity, enthusiasm and dedication became quickly evident on the occasion of our first meeting

Sheri was among a number of specially invited guests (key to the food service industry in Victoria) at the University of Victoria when the then newly-published FOODSAFE course was introduced in 1988.

It became apparent during the lunch break that day that Sheri was going to "coat-tail" me for the rest of the session. My lunch remained mostly uneaten. Chewing was a luxury Sheri wouldn't allow me to indulge. Every bite was accompanied by a battery of questions and comments that demanded an immediate response. She made it perfectly clear that she believed the food industry needed the new FOODSAFE training program. I believed it, too.

Due to the dedication of Sheri Nielson and a core of outstanding FOOD-SAFE instructors, this program has become an overwhelming success in British Columbia, exceeding all expectations. FOODSAFE has also become accepted as standard for several international events and jurisdictions.

Sheri's insight for her industry is right on track, and her anticipation of education as a risk management tool for foodborne illness prevention was soon adapted as an ideology for the domestic kitchen chef.

Sheri is a treasure to the food service industry, to us in the health protection field, and to you; the home chef.

Sanitary food handling practices work. Sheri knows it, and I know it.

It works in the industry and it will work for you in your home—if you observe and practice the principles and rules presented in this book.

This book is written so that technical and scientific terms are kept to a minimum. However, some references to science are unavoidable. A glossary of terms is provided for your easy reference.

Microorganisms have been referred to by their "common" names in the text; their proper names appear in the Glossary and Sources of Foodborne Illness. To the scientific purist may I say thank you for your patience.

Wm (Bill) Hines, C.P.H.I. (c)
Food Safety Consultant,
Registered Environmental Health Officer

Introduction

Jamie lovingly approached the task; her first homemade soup. The kind she remembered grandma preparing for the men when they came in from the fields for lunch.

First, a big, meaty soup bone and some stew meat. Cover with water and bring to a boil. Add salt and pepper and a couple of bay leaves. Mmm, the smells were great.

Then she lowered the temperature, chopped up some carrots, cabbage, celery and a big onion which she added as the mixture simmered. After a couple of hours she removed the bones and cut off the meat, now tender, into bite-sized chunks. She had too much meat so she decided to set some aside. While she finished preparing the soup, she allowed the extra meat to cool then put it in the refrigerator. Maybe she would add it later.

Next, some tomatoes were mixed in, a bouillon cube for added flavor, a touch of Worcestershire sauce and a sprinkling of Oregano. There was still time for another hour of simmering before Byron came home from work. She hoped he'd like it. It was a neat thing to do on her day off. Apart from cutting her finger while chopping the vegetables, she'd really enjoyed the experience. Sort of old-fashioned homemaking.

The soup was marvellous. Byron had seconds and later had a small third helping. He even expressed a couple of compliments while perusing the stock market quotations. And there was enough of the delicious blend left over for lunch the next day. Jamie poured it piping hot into Byron's big thermos and put it in his briefcase before they settled in for a quiet night of TV. The next morning, she remembered the extra meat in the refrigerator and added it to the thermos before replacing it in Byron's briefcase.

The soup was only lukewarm when Byron sat down to lunch, but no problem—just nuke it in the office microwave and he had another great batch of steaming hot soup that afternoon.

By three o'clock that afternoon, he had severe stomach cramps. By four, he was nauseous and vomiting. His agonizing condition, later accompanied by diarrhea, lasted for several hours. A visit to the doctor and some testing confirmed that he had suffered from "staph" poisoning caused by *Staphylococcus* bacteria.

An isolated case? Not at all. Every day throughout North America, there are hundreds of similar cases. Most are not life-threatening. Many are mistakenly diagnosed as "24 hour flu." But,they are harmful; at best very unpleasant—and in the case of the very young, the elderly, or people who are already weakened by some other illness, they can be fatal.

The case of Jamie's soup will be analyzed later, but first some background music.

In the days when grandma ruled the kitchen and perhaps unwillingly, spent most of her life there, kitchen lore—including menus, recipes, kitchen management and sanitary practices—was handed down usually from mother to daughter in a natural and orderly manner. The kitchen was a classroom where the daily routines of planning, shopping, food storage, meal preparation, cleaning and sanitation were absorbed as readily as learning to talk or tie one's shoes.

That is not to suggest that all situations were ideal or that mistakes were not made, even in the best-managed kitchens. But, there was a body of folk knowledge which tended to be passed on from generation to generation. Interestingly, some parts of this reservoir of information have not only survived but have been enlarged and enhanced. A plethora of cookbooks and dietary publications has given us a rich literature of menus, recipes and culinary techniques. But how many of these books describe safe procedures for thawing frozen foods? How many recipes include advice on hot holding temperatures or the sanitizing of cutting boards after cutting raw meats? Do any of them, in fact, tell us why certain foods must be cooked to the correct temperatures?

We don't want to create the impression that sickness and death are waiting around the corner each time you attempt to fry an egg or prepare a Thanksgiving turkey dinner. At the same time, every home would be a safer place to eat if the basic principles of sanitary food handling were understood and practiced by all family members.

Times have changed since that somewhat idyllic scene described in grandma's kitchen. Today many homes have no one presiding over the kitchen on a regular basis. Single-parent and double-income families are now the rule, rather than the exception.

Kitchen duties are more often rushed and shared among family members. There is little time for the preparation of large, many-course "old-fashioned" meals. The emphasis has shifted to convenience foods requiring a minimum of planning and preparation. Technological change has aided this transition with the development of modern and efficient kitchen appliances. Freezers, refrigerators, convection ovens and microwaves, dishwashers, garburetors, better ranges and self-cleaning ovens have reduced the kitchen chores with the added bonus of improved food sanitation capabilities.

But, while a new lore has grown around the changes, most of it is dedicated to the taste and appearance of the food rather than health risks and safety. Many people don't take the time to learn how to use the new appliances safely and effectively. (Quickly! How long can you safely keep a piece of leftover roast beef in the refrigerator?)

Advances in food manufacturing, processing, marketing and distribution systems may be having an even more profound effect on home cooking. Giant supermarkets, thousands of new products and a myriad of packaging techniques—not to mention long lineups at the cashier's wicket—confront the harried shopper trying to grab a few "vittles" on the way home from work. It is difficult, in the best of situations, to discover where the products have come from, let alone how they were packed (frozen, refrigerated, vacuum-packed, bottled, tinned, pasteurized, freeze-dried, dehydrated).

And now that you have them home you have to read the labels carefully to determine where and how to store and prepare them. While there is no doubt that these processes reduce human contact with food and represent improvements in sanitary control, they tend to be confusing to us and may lull the unsuspecting buyer into the assumption that all these products are completely safe. Many an amateur chef has opened one of these packages to discover that the food has spoiled. Worst of all, some contaminated foods may display no tell-tale signs such as a disagreeable odor or discoloration. Many an unsuspecting family has eaten risky food.

Whereas incidents of contaminated food supplies could once be traced to local sources, it is now possible for an error in, say, the manufacture of cheese or the application of a pesticide to result in the distribution of contaminated foods all over the world. Strict regulation and vigilant enforcement minimize the opportunity for this type of problem in many countries, but with the enormous volumes, vast distances, and intricate distribution networks, it is almost inevitable that problems will occur.

Scary stuff? Maybe, but it doesn't have to be. A working knowledge of the causes of food-related illness and the conditions which influence them will make the task of safeguarding your family's food much easier. The control and regulations governing agriculture and food and drugs are continually improving. The workers throughout the food industry are provided with more opportunity for quality assurance and training, and the facilities for testing and processing and equipment are getting better every year.

The weak link in the chain is us—the home chef.

It's a circumstance which can be improved by the application of a few simple guidelines and easy-to-follow safeguards. The pages that follow contain useful background information, basic principles, simple "rules of thumb," checklists and many practical examples which will contribute to reducing foodborne illness and minimizing the risks in your home.

Now, what about the problem of Jamie's soup? To begin with, we can't prove the soup was the source of Byron's illness. That could only be accomplished if a sample of the food was thoroughly tested in a laboratory and matched resulting samples with lab-confirmed specimens from Byron. This, incidentally, is the reason health officers often have to make an educated guess at the actual cause of food poisoning. In many cases, the food has either been consumed or thrown out before the illness can be identified or specimens from the victim were not taken for analysis.

But to return to the soup: we can make a pretty good case for it as the source of discomfiture. We believe the soup was prepared carefully and correctly. Good personal hygiene and sanitary practices were almost practiced throughout its preparation. Correct cooking temperatures were used in the boiling and simmering stages. The soup was served piping hot. Neither of them got sick after the evening meal, despite the consumption of large helpings.

The problems occurred after the evening meal. Remember that the soup was placed in the thermos at night which would have allowed it time to cool somewhat, in the thermos, by morning. Then, in the morning Jamie added some of the extra meat which had been in the refrigerator all night. The effect of this would be to further cool the soup in the thermos to what is often called the "Danger Zone," between 40°F (4°C) and 140°F (60°C). This is a temperature range in which harmful bacteria are known to reproduce rapidly.

But where did they come from? Probably from Jamie's hands, in the morning, when she cut up the extra meat from the refrigerator. She had cut her finger the day before and it may have started to become infected by morning. The same bacteria which cause "staph" poisoning are commonly responsible for infection in cuts, sores, and pimples. It is also possible that she coughed or sneezed on the meat while putting it in the thermos, or she forgot to wash her hands after putting her cat, Fluffy, out in the morning. Approximately 50 per cent or more of all humans are estimated to carry *Staphylococcus* bacteria in their noses and mouths or on their normal skin. And who can tell about Fluffy?

So the bacteria were there. They are everywhere! There was lots of food and moisture for them to grow, and the temperature in the thermos was just right for that development. All the conditions were right for a foodborne illness to happen. It would be a few hours before Byron opened his thermos for a delicious, but unhealthy surprise—plenty of time for the bacteria to multiply.

But, wait a minute. Didn't he put the soup in the microwave? And wouldn't that destroy the bacteria? Yes, it would! But, unfortunately this particular bacterium produces a toxin or poison as a waste product as it grows. It is colorless and odorless, but it can make you very sick. The heating in the microwave would destroy the bacteria, but not the toxins they produced.

There are several lessons in this story, but the most important ones are sanitation and temperature control. Had the soup and the extra meat been hot enough in the thermos and not cross-contaminated by unclean hands or an infected cut, the bacteria would not have had time or opportunity to multiply and produce their harmful toxins.

Which brings us to the other question. How long can you safely leave cooked beef in the refrigerator? If you replied "as long as you like, provided you don't eat it," you deserve half marks and a scholarship to the Groucho Marx school of humor. But, if your answer was any more than two to three days, your next trip may be to the doctor's office. Bacteria may survive and even multiply at refrigerator temperatures (less than 40°F or 4°C), albeit much more slowly than at room temperature.

By the way, don't worry if you can't remember all of this. That's why you bought this book.

— Barry Black

Microbiology

Microbiology is probably the last subject a homemaker thought he or she would need to study. Gardening or home economics would seem to be more sensible topics.

It's easy to see how foods contaminated with pesticides or chemicals can cause illness, but food specialists tell us biological contamination is the main cause of food poisoning.

The soup-making episode described in Barry Black's introduction involved bacteria being allowed to multiply in food. Most food poisonings are caused by bacteria, but other microbes—viruses and parasites, for instance—can also be carried and spread in food. When we understand how these microbes survive and multiply, we can develop strategies to control or destroy them. Only then can we be sure we are serving our families safe meals and snacks.

The good news is it's not difficult to protect ourselves and the foods we eat, provided we learn a few important principles and follow some basic rules.

Welcome to the world of microbiology

What are microbes?

First, we need to realize that microbes are alive. The tiniest organisms in the animal kingdom, invisible to the naked eye, can be observed only under a microscope. They're in air, in soil, in water, on dust, on insects and on animals, on humans and even on each other. That means they're everywhere, including most foods.

We live with microbes every day and most are harmless. Some are even helpful: like the bacteria used in cultures to make cheese and yogurt, the yeasts used in breads, and the molds used to produce certain cheeses and antibiotics that are used as medicines.

Certain microbes, such as bacteria and molds, will form spores when conditions for their growth are not right. Spores act like tiny seeds that remain dormant until conditions for revival are present. Then, they can emerge as bacteria or mold which multiply quickly.

About one per cent of microbes are harmful to humans. Known as pathogens, these are the microbes we need to control or destroy in our foods and kitchens to prevent food poisoning.

It may help to think of microbes in terms of "The Good," "The Bad" and "The Ugly." The good ones are helpful to us in making foods and medicines, the bad ones cause food spoilage and the ugly ones are pathogenic and make us sick or worse.

How do microbes get around?

Microbes don't have legs, fins or wings, so they have difficulty moving around on their own. They relocate from one place to another by "hitch-hiking." They can catch a ride on or in people (Measles), animals and birds (*Salmonella*), pets (*Campylobacter*), insects (Malaria) and water (Cholera), and they can be carried on air currents and dust.

In our kitchens, microbes can hitch-hike from a cutting board that is used to cut up a raw food like chicken to many places. For example, if there was *Salmonella* bacteria on the chicken they can be picked up on the cloth we use to wipe the board and our hands when we touch the chicken. They can be carried to the counter top if we wipe it with the same cloth, to the sink where the cloth is rinsed and to the taps that are turned on for rinsing, and they can be on the knife. This is often referred to as cross-contamination.

Bacteria

Pathogenic bacteria cause the majority of microbial food poisonings. These critters are very different from us in the way they reproduce. They don't have to grow up, go out on dates, fall in love and be married before they begin a family.

They grow by binary fission, or more simply put, by dividing into two identical organisms. Under the right conditions, this can happen every 10 to 20 minutes. So two become four, four become 8, 8 become 16, and so on. In two hours enough harmful bacteria could grow in food to make a person ill after eating it.

Binary fission

There are two points to remember about bacteria. One, they don't grow in size, getting bigger all the while—they grow in numbers. Big numbers! Two, bacteria can survive in a wide range of temperatures and environments from hot springs to the arctic.

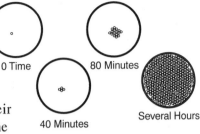

0 Time   80 Minutes

40 Minutes   Several Hours

Also, remember that certain bacteria can develop spores if conditions for growth are not to their liking. They can then emerge from the spores and grow when conditions improve, even after many years and under the harshest conditions.

Imagine our bacterial spore as a single dried bean. When the bean was harvested it contained a small amount of water, like a bacterial cell. During processing, the water was removed to allow the seed to be safely harvested and stored for some time (like a spore).

When the bean is soaked in water to make baked beans, the seed absorbs water and becomes more like a fresh bean than a dry seed. In fact, if you soak it long enough under the right conditions, or plant it, it will sprout! The seed (bean or spore) contains all the genetic information it needs to become a new plant or bacterial cell.

Bacteria are known to have survived freezing for as long as 135 years. They've also caused many deaths arising from diseases such as Cholera, Tuberculosis and Botulism.

We can concentrate our efforts against microbes by dealing with the pathogenic bacteria. If we learn how to control them, we can then use the same methods to control the dangerous viruses and parasites associated with our food.

Viruses

Viruses are the very tiniest of organisms that can survive and be carried in food. Viruses that cause disease in humans cannot grow in food, but they can use it as a vehicle of transmission. Once they enter the human body, they may cause diseases like Hepatitis and Measles, as well as colds and flu.

Fungi (yeasts and molds)

Certain fungi are helpful to us, while others are pathogenic and may make us ill when we eat them or their by-products.

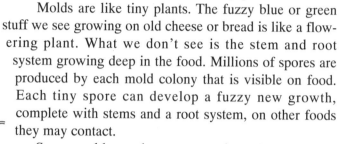

Molds are like tiny plants. The fuzzy blue or green stuff we see growing on old cheese or bread is like a flowering plant. What we don't see is the stem and root system growing deep in the food. Millions of spores are produced by each mold colony that is visible on food. Each tiny spore can develop a fuzzy new growth, complete with stems and a root system, on other foods they may contact.

Some molds produce mycotoxins which become part of the food. Some toxins cause illness immediately while others are thought to be carcinogenic or mutagenic.

Mold & its spores

Parasites

Parasites include tapeworms, roundworms and other ugly things sometimes found in soil, fish, meat and surface water. Some parasites can be transferred to food from unclean hands or through cross-contamination, but the main vehicles are inadequate cooking of certain foods and drinking untreated water.

Protozoa

Protozoa are tiny, simple, one-celled animals—like amoebas—that can also be parasitic. They can be found in soil or in the intestines of man and animals. Most protozoa are harmless to us and actually eat bacteria as a main course. *Giardia*, which cause the disease known as Beaver Fever, is an example of a pathogenic protozoa that can be carried in food and untreated water. The tropical disease, Malaria, is another example of a pathogenic protozoa.

Toxins

Certain microbes produce toxins or poisons that are formed in foods as they grow, multiply and die. If these microbes grow on or in foods, the foods can become contaminated with the toxin.

Some of these toxins are considered to be heat stable. Even when foods are heated to temperatures much higher than cooking temperatures the toxins are not destroyed.

What do microbes need?

There are certain things bacteria need in order to survive and multiply. Many of these things (other than falling in love) are the same ones we need to survive.

Food

Like all living things, microbes need food. They prefer foods that are rich in protein, carbohydrates, vitamins and are moist, just like many of our favorite foods.

Certain foods are referred to as "potentially hazardous" because they have the characteristics that microbes prefer. These potentially hazardous foods include: meat, poultry, seafood, dairy products, raw eggs, cooked cereals, cooked rice and cooked vegetables.

Temperature

Pathogenic bacteria grow most quickly at temperatures near human body temperature of 98.6°F (37°C). They would grow very quickly on food in your kitchen on a hot summer day, but they also get along quite comfortably anywhere between 40°F (4°C) and 140°F (60°C)—a temperature range from inside the refrigerator to very hot water from a faucet.

Body Temperature 98.6°F (37°C)

Hot Soup 140°F (60°C)

Danger Zone

Refrigerator 40°F (4°C)

Pathogenic bacteria do not grow as quickly at the extremes of this temperature range, but they can still grow and be a danger to us.

Moisture

In order to use food, bacteria need moisture that is readily available. Since microbes don't have teeth, they need moist food so that they can absorb the nutrients dissolved in the water.

Oxygen

Just like us, molds, protozoa and some bacteria need oxygen to live. Others only grow when oxygen is not present. The bacteria responsible for Botulism are a good example of microbes that grow where oxygen is absent. That's why they can pose a hazard in home-canned foods that have not been properly processed.

Many bacteria, particularly the pathogenic bacteria, can grow in either the presence or absence of oxygen.

As you can see, most microbes are very versatile. Just like us, they can thrive in a variety of environmental conditions.

pH

pH is a value expressing the relative acidity or alkalinity (base) of a substance. pH is expressed as a number on the scale of zero to 14. Zero to seven is acidic, seven is neutral, and seven to 14 is alkaline. Zero is most acidic, and 14 is most alkaline.

Microbes prefer neutral or slightly acidic foods. Creamy sauces, high protein foods, soups and gravies are some good examples of neutral pH foods.

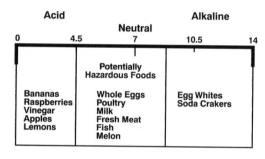

pH scale

Time

It takes time for enough bacteria to grow in food and cause food poisoning. However, growth can be rapid under the right conditions. In a few hours, a single bacterium can produce billions of cells just like itself. But, you never find just one bacterium on food. In fact, there may be millions of bacteria to start with, so population explosions are common.

How can we destroy or control microbes?

The good news is we don't need different means to control each different type of microbe. All of the dangerous microbes will be controlled by applying the following principles.

Temperature

Temperature is your most powerful tool in destroying or slowing the growth of microbes.

High temperatures for sufficient periods of time will destroy the microbes that can grow in food and make people ill. If you heat food to 170°F (77°C) for several minutes, you will destroy any disease-causing microbes in the food.

But, how can we even think of cooking a medium-rare prime rib at this temperature? Then, time becomes our ally. With correct cooking temperatures for the correct amount of time, we can destroy pathogens on the surface of a whole cut of meat provided that it has been properly prepared.

It was thought by some that freezing temperatures would eventually kill microbes. This is not always the case, as many pathogenic microbes survive freezing.

By controlling temperatures, we control how quickly microbes will grow in numbers or die.

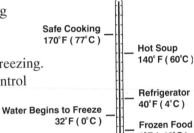

Safe Cooking
170°F (77°C)

Hot Soup
140° F (60°C)

Refrigerator
40°F (4°C)

Water Begins to Freeze
32°F (0°C)

Frozen Food
0°F (-18°C)

Kitchen Sanitization

Contrary to popular belief, a squeaky clean kitchen won't, by itself, prevent food poisoning. It certainly helps, but we must remember we are dealing with the world of pathogenic microbiology—an unseen but ever present danger.

Here, we must again use our imagination. Our kitchen is "squeaky clean" and we're cutting beef steak into cubes to make kabobs for the barbecue. We now know there are microbes "out to get us" on the knife, the cutting board, our hands, and perhaps the counter top. To rid ourselves of these microbes, we must sanitize the contaminated surfaces.

Sanitizing is a three-step process: We have to always make things squeaky clean again before sanitizing. So, **step one** is washing the counter top, cutting board, knife and our hands with warm soapy water before we sanitize. **Step two** is rinsing them to remove any soap or residue.

That little scratch on the cutting board appears to be only a tiny scratch to us—but for a pathogenic enemy, it's a Grand Canyon to hide in. Washing up with a detergent and water followed by a clean water rinse gets rid of the cover the cell can hide in.

Step three—the sanitizer—is terminating the cell.

Several chemicals are available to sanitize work surfaces and utensils. You can use a solution of household chlorine bleach or iodine added to water to destroy microbes in the kitchen. Because it stains surfaces, iodine is not as practical as household bleach.

A good rule of thumb is one ounce (30 ml) of bleach in a gallon (4 L) of water. By the way, this is the common sanitizer used by professional food handlers and chefs. When used in this fashion, bleach is much less harmful to the environment and to us.

In fact, it works like magic. Once a surface or utensil has been properly sanitized and dried, no chemical residue remains. It magically evaporates like water, leaving the item pathogen-free.

Personal Hygiene

The food handler has been found to be the source of contamination in the majority of foodborne illness cases. If you are the person cooking, then this means that you could be the source of contamination unless you pay close attention to your personal hygiene.

Because we touch food with our hands during preparation, **hands** can become the source of pathogen transmission. Microbes may come from our bodies, as with *Staphylococcus* bacteria, or they may be on our hands because of cross-contamination after handling raw foods such as chicken that contained *Salmonella* bacteria.

This is also why it is recommended that, where possible, we use clean utensils rather than our hands when we prepare food. If we reduce hand contact, then we also reduce the chance of transmitting pathogens to the food.

Learning and practicing good personal hygiene is a very important step in the safe preparation of food. The simple practice of frequent hand washing before and during food preparation can significantly reduce the risk of contamination of the food.

Food Protection

Chicken and Poultry

Poultry is popular around the world. We have become quite proficient at producing, transporting and preparing this commodity in countless ways. If we remember that microbes get around by "hitch-hiking," we can see how it has been possible for microbes such as *Salmonella* and *Campylobacter* bacteria to cause food illness everywhere.

Because *Salmonella* and *Campylobacter* are commonly found on and in chicken and other poultry, it makes sense that these bacteria have taken the opportunity to flourish wherever poultry is a popular part of the human diet.

The *Salmonella* commonly found in chicken does not normally cause illness in the bird or fowl. It's people who get sick from consuming this type of bacteria.

Because so many poultry products have been found to be contaminated with *Salmonella*, it has become necessary to handle all poultry as if it is contaminated.

The bacteria live in the innards and on the skin and feathers of the bird. They can be spread to the meat during processing.

Campylobacter and other pathogenic microbes can also be a danger in raw chicken and poultry. Bacteria find many of the things they need for survival in raw fowl—a soft, moist, protein-rich and neutral pH food. By controlling temperature, we can control or eliminate all pathogenic microbes in these foods, ensuring our turkey or duck dinners are safe to eat.

You can be assured that *Salmonella* and other bacteria have been destroyed when the poultry reaches an internal temperature of 170°F (77°C). Because of the varied thickness resulting from the shape of whole poultry items like chicken, it is difficult to be certain that the heat has actually penetrated all parts of the bird. **That is why it is recommended that whole birds be brought to in internal temperature of 180°F (82°C).**

Bacterial growth will be slowed at refrigeration temperatures of 40°F (4°C) or colder. Keep in mind, refrigeration is not lethal to these microbes, but keeping them in a cold environment helps control their rate of growth.

Purchasing

▶ Always buy chicken and poultry from a reputable source.

▶ Check the thermometer in the store's display case. Chicken and poultry should always be held and displayed below 40°F (4°C). If the cooler is not cold enough, there could already be countless microbes growing in the food.

▶ Check the date the product was packaged. Poultry should be cooked or frozen within three days of processing. If it's been in the display cooler for two days, you need to cook it within one day. It's always best to buy the freshest product possible.

▶ Darkening of the wing tips is a sign the chicken has been sitting on display, for some time.

▶ An off odor under the wings or between the thigh and the body indicates the poultry is spoiling. There should be little or no discernable odor if the product is fresh.

▶ Frozen poultry should be thoroughly frozen. If there is moisture on the package or frost inside the package, the freezer may not be working properly and the food may have been partly thawed and refrozen.

▶ Raw poultry that has been frozen must be cooked thoroughly before it can be safely refrozen.

Storing

▶ Poultry must be frozen or held at refrigeration temperatures of 40°F (4°C) or colder until it is cooked.

▶ Frozen poultry must be held at -0°F (-18°C) or colder.

▶ Raw poultry should not be allowed to contaminate other foods by touching or dripping onto them. Store the raw poultry on a low shelf in a container that will not leak onto ready-to-eat foods.

▶ It is important to refrigerate poultry as soon as possible after it is purchased. Foods which most readily support the growth of pathogens should be purchased at the end of a shopping trip and placed in cold storage immediately.

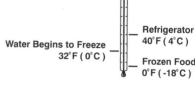

Water Begins to Freeze
32°F (0°C)

Refrigerator
40°F (4°C)

Frozen Food
0°F (-18°C)

Thawing
- Potentially hazardous foods like frozen poultry are best thawed in the refrigerator. This may take two days and even longer for large birds such as turkeys, so planning ahead is necessary. Safe thawing alternatives include microwave ovens, which can do an excellent job if they have a good defrost setting, or running cool water over the plastic-wrapped item, followed by immediate cooking.
- NEVER thaw frozen poultry at room temperature. Foods that have been frozen are easier for bacteria to use; they can multiply rapidly on the warmer outer surface of foods that are still frozen inside.

Preparation

NOTE: any surface that comes in contact with the poultry or the drippings can be contaminated by hitch-hiking or bathing microbes.

- To prevent cross-contamination, keep a separate work area or cutting boards and tools for preparing or handling raw poultry. This also makes it easier to clean up all possible contact surfaces when you've finished working.
- After you finish preparing the raw poultry, all work surfaces and tools should be washed and then sanitized using a solution of water and chlorine bleach. The recommended ratio is one ounce (30 ml) of bleach per gallon (4 L) of water.
- If your recipe requires a marinating period, it should always be done in the refrigerator. Some cooks prefer to leave the item marinating for a longer time when it is refrigerated.
- Don't forget to wash your hands after handling raw poultry.

Cooking

A metal probe thermometer is one of the very best investments you will ever make to ensure the health of your
family. This inexpensive item can be found in cooking specialty shops or grocery stores. They're
available in dial or digital types. Since
they work on a bimetallic coil principle there is no danger from mercury
poisoning or contamination if you break one of these essential
kitchen gadgets. Don't forget to sanitize it before you use it.

- ► To kill any microbes in the food, **whole poultry** must be
  cooked to an internal temperature of 180°F (82°C) for a few
  minutes. NOTE: This is not the temperature of the oven—
  it's the temperature of the meat deep inside the bird.
- ► When checking the temperature with a probe thermometer,
  check the thickest part of the meat. The tip must not touch
  the bone. The bone has a different density and conducts heat
  more quickly than the meat so your reading will not be accurate.
- ► Cook breasts or large thighs of poultry until your probe thermometer registers a safe internal temperature of 170°F
  (77°C).
- ► It is safe to assume that wings, legs and ground poultry have
  reached a safe temperature of 170°F (77°C) when the juices
  run clear and there is no pink inside.
- ► Some famous recipes tell us to cook poultry to a pink stage.
  Because *Salmonella* and other
  bacteria may still be alive at this
  stage, this is not a safe recipe.
- ► Don't leave the thermometer in the
  food while it is cooking. Use it
  when you are checking for "doneness."
  Because the metal probe or glass can pick up the
  temperature of the air in the oven, the accuracy of
  your reading can be affected. Probe thermometers
  are not designed to reach high oven temperatures,
  and prolonged exposure could blow off the top of
  your thermometer.

Poultry Cooking
180°F (82°C)

Safe Cooking
170°F (77°C)

Hot Soup
140° F (60°C)

► Partial cooking followed by a finishing stage at a later time is a risky practice. Partially cooked poultry may feel hot on the surface when we touch it. But, in fact, it means we have warmed the inside of it to the microbes' favorite temperature and softened it making it easier for them to use. The partial cooking and cooling process allows microbes time to grow to dangerous levels.

► If you want to baste poultry with a leftover marinade, the marinade should first be brought to a boil to destroy any pathogens that may have been in the raw food. It is important to protect your cooked food from contamination by raw food.

► If you pre-cook poultry to finish on the barbecue, grill it immediately. Don't let it "rest" in the danger zone temperature range.

Cooling, Storage and Reheating

► Potentially hazardous foods like cooked poultry must be stored in the refrigerator or freezer as soon as possible after cooking. If these foods have been allowed to remain in the danger zone for more than two hours, they should be thrown out. So, rather than waste these foods, make sure you refrigerate them promptly.

► Refrigerated leftovers should be used within three days and consumed cold out of the refrigerator or heated to piping hot temperatures. Never just "warm up" leftover poultry and then eat it. It could be hazardous to your health.

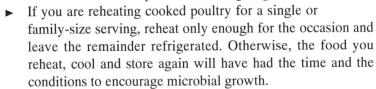

Hot Soup
140° F (60°C)

Body Temperature
98.6°F (37°C)

Danger Zone

Refrigerator
40°F (4°C)

► Make sure cooked poultry is reheated quickly to a piping hot 170°F (77°C). That will destroy any microbes that have been introduced since the previous cooking stage.

► If you are reheating cooked poultry for a single or family-size serving, reheat only enough for the occasion and leave the remainder refrigerated. Otherwise, the food you reheat, cool and store again will have had the time and the conditions to encourage microbial growth.

► Leftover poultry gravy should be brought to a boil before it is served. It should be frozen or reheated only once.

Stuffing Large Poultry

► It can take over two hours for the heat of the oven to penetrate a large stuffed bird. Any bacteria present in the dressing will have the time, temperature and nutrients to grow to dangerous levels. If those bacteria produce toxins, the toxins will not be destroyed, even when the temperature is hot enough to kill the bacteria. This means the food can contain enough toxins to make someone ill.

► When a large poultry item has been frozen, there may still be some ice crystals in the cavity when it is stuffed. The stuffing would stay in the danger zone temperature range even longer than it would in a completely thawed bird.

► **Stuffing should be cooked separately.** The flavor of the stuffing can be enhanced by basting it with the cooked juices from the bird.

► If tradition demands that you must cook a predressed bird, it is safer to use smaller birds, because the heat can reach the stuffing sooner. Stuffing the bird immediately before putting it into a pre-heated oven can help reduce the time the stuffing spends in the danger zone temperature range. Make sure the internal temperature of the bird reaches 180°F (82°C), and the stuffing reaches 170°F (77°C) before it is served.

► Remove the stuffing from the bird as soon as possible and refrigerate it in a separate container. Stuffing can be dense and may take too long to cool if it is left in the cavity.

Barbecue Safety

► Keep raw chicken and poultry at refrigeration temperatures of 40°F (4°C) until it's time to grill them.

► If the pieces of poultry you are about to cook are too large for the heat of the grill to penetrate through the meat without burning the outer surface, the item should be precooked and immediately finished on the barbecue.

► If you want to baste the poultry with a leftover marinade, the marinade should first be brought to a boil to destroy any pathogens that may have been in the raw food. It is important to protect cooked food from contamination by raw food.

► If you use a brush to baste raw poultry at the beginning of your cooking period, the brush becomes contaminated with the pathogens on the surface of the raw foods. If you baste the cooked, ready-to-eat food with that same brush, you will recontaminate the cooked food with the pathogens you picked up on the brush.

► The cooked poultry can be contaminated by drippings on the platter used to carry the raw foods to the barbecue. Use a clean platter when taking cooked foods off the grill.

Poultry Handling — Myths and Facts

Myth: Marinated chicken is free of microbes because they are killed by the alcohol in wine or brandy and the neutral pH that bacteria thrive in is altered by adding lemon juice or spices.

Fact: Alcohol is not an efficient sanitizer to begin with. Both the alcohol and the acid in lemon juice are diluted by the juices from the raw chicken and so have no real effect on the neutral pH of the food. Spices like cayenne or curry taste hot to us, but have no influence on microbes. The only "heat" that will destroy them is thermal heat of 170°F (77°C).

Myth: Free range chickens don't have Salmonella because they are raised "organically."

Fact: Salmonella bacteria can grow anywhere in the world. They don't depend on a specific climate or environment to be present. They could be found in chicken feed or passed to future generations in the embryo.

Meats

The meat from animals is high in vitamins, minerals, moisture and has a neutral pH level. These foods offer a perfect environment for any bacteria present to grow rapidly when they remain in the danger zone temperature range. Meat can be contaminated with microbes from the innards or the hides of animals because they may be transferred to the meat during processing.

Beef

Beef is frequently contaminated with *E. coli* bacteria. It can be found on the outer surface of a cut of meat. This occurs during processing and is difficult to prevent. Beef may also be contaminated with other pathogens like *Salmonella* or parasites. It is important that any outer surface of the meat is heated to the temperature that will kill dangerous microbes.

Bacteria don't have teeth. They cannot chew their way through a cut of meat so they cannot get inside a **roast or steak** unless we let them hitch-hike on a fork, knife, thermometer or grinder. Using tongs or lifters keeps dangerous bacteria on the outside of the meat until they are destroyed by the heat of cooking. The heat in the oven, on the grill or in the frying pan will be high enough to destroy any bacteria on the surface. That is why medium-rare roast beef or steaks are of less concern, provided they have been handled and stored properly.

Ground beef is an entirely different story and is particularly hazardous because it is made up of many, many surfaces—each of which may carry bacteria. Ground beef can cause serious illness—even death—from *E. coli* food poisoning if it's not thoroughly cooked to the temperature that will kill pathogenic bacteria. Since it takes very few of these particular organisms to make someone seriously ill, it is important to make certain that all parts of the meat reach the correct temperature. To destroy any bacteria that may be in the meat, ground beef should be cooked to a well done stage.

Roasts that have been **boned, rolled or stuffed** may have microbes at the center because the butcher's knife or skewer could transport them inside from the outer surface. These cuts of meat should be cooked thoroughly to ensure all harmful microbes are destroyed.

Diced or chopped beef can have microbes on all surface areas. It's important to make sure that all sides receive adequate heat to kill any pathogens.

Steak tartar and other recipes that use uncooked meat are a serious risk to your health and are not recommended. Only a high heat will destroy the dangerous microbes that we must expect to find on beef.

Beef can also be—and often is—contaminated with other microbes, such as *Salmonella* and *Clostridium perfringens*. These microbes will be destroyed at correct cooking temperatures.

Eating uncooked meat is not recommended

Pork

There has been a concerted effort over the past several decades to eradicate *Trichinella* parasites and their eggs from the flesh of pork. These efforts have been quite successful at government inspected hog farms. Pigs from farms which are not subject to government inspections are more likely to be infected with *Trichinella*. It is recommended that pork purchased directly from the farm be thoroughly cooked. We cannot tell if pork is infected just by looking at it so you may chose to handle all pork as though there is a risk of parasitic infection.

The *Trichinella* parasite finds a host (a pig, bear or walrus), and migrates through the stomach to the muscles where it lays its eggs. It then waits for another host (us) to eat it and is transferred to the new host unless it is first destroyed through proper cooking.

Pork has also been found to be contaminated with *Campylobacter* bacteria. It has been associated with *Toxoplasma*, another parasite that is particularly hazardous to pregnant women and their fetuses.

As an extra precaution, pork can be frozen at 0°F (-18°C) in a deep freezer to destroy parasites and their eggs. Cuts that are up to six inches (15 cm) in diameter should be frozen for 20 days while cuts from six inches to 26 inches (15 cm to 68 cm) in diameter should be frozen for 30 days.

Lamb

Lamb may also be contaminated with pathogenic bacteria which can be found on the outer surface of a cut of meat. This occurs during processing and is difficult to prevent. It's important that the outer surface of the meat is heated to the temperature that will kill dangerous bacteria.

Bacteria don't have teeth. They cannot chew their way through a cut of meat so they cannot get inside a **roast or chop** unless we let them hitch-hike on a fork, knife, thermometer or grinder. Using tongs or lifters keeps dangerous bacteria on the outside of the meat until they are destroyed by the heat of cooking. The heat in the oven, on the grill or in the frying pan will be high enough to destroy any bacteria on the surface. That's why medium-rare roast leg of lamb or chops are of less concern, provided they have been handled and stored properly.

Wild Game

Some people think of wild animals as being "pure" and don't realize their meat can be contaminated. What we may not understand is that microbes occur "naturally" in our environment, and that includes the great outdoors.

The same microbes and parasites that find our domestic animals appealing can be present in any wild game. Meat from wild animals should be cooked to a well done stage.

As an extra precaution, all game meat and especially bear, wild boar or walrus should be frozen at 0°F (-18°C) in a deep freezer to destroy parasites and their eggs. Cuts that are up to six inches (15 cm) in diameter should be frozen for 20 days while cuts from six inches to 26 inches (15 cm to 68 cm) in diameter should be frozen for 30 days.

There is a strain of *Trichinella* that survives freezing in meat raised in cold climates like the arctic region. This strain has developed resistance to freezing temperatures and so the parasites will not be destroyed through freezing. Meat from the arctic regions **must** be thoroughly cooked to a well done stage to destroy these parasites.

Safe Cooking 170°F (77°C)

Boiling Water 212° F (100°C)

Hot Soup 140° F (60°C)

Bacon, Ham and Sausages

The processing of these foods has changed in important ways over the years. Consumers have demanded foods with less salt and fewer preservatives. The result is that foods which traditionally have been relatively safe from rapid spoilage or contamination by pathogens are now potentially hazardous in the same way as unprocessed meats.

As consumers, we have come to think that certain descriptions like "smoked" means "preserved." This is not the case in many foods described as "smoked" today.

In the past, meat and fish were preserved by being dried or salted. The process took time and smoke was kept on the food to keep insects off while the drying continued. By drying and salting the food, the pathogens did not have one of the conditions they need to survive—moisture. That, and deposits from the smoke, is what preserved them. But, all too often some spores survived, and Botulism or "sausage poisoning" occurred.

Today, most of the foods described as smoked have simply had the flavor of smoke added to them. Bacon, ham and sausage must be handled the same way as any other meat unless they have been thoroughly dried—like beef jerky, or cured with salt and a reduced moisture—like salt herring.

If the food has been processed in such a way that it is safe for a long period of time or without refrigeration, it will be indicated on the manufacturers label.

Sometimes these foods have a "use by" date on the package. This means the food will be safe in the unopened package until that date if it is stored properly. Once the package is opened, the food should be used within three days.

Delicatessen Meats

Because many delicatessen style meats have been cured or flavored to some degree, we may mistakenly think they are somehow preserved. These foods need to be handled the same way as all cooked meats. They need to be refrigerated and should be used within three to four days after purchase.

Purchasing Meats

► Always buy meats from a reputable source.

► Check the temperature in the display case. Meats should always be held below 40°F (4°C). If the cooler is not cold enough, there could be an unacceptable number of pathogens in the food and it may have begun to spoil.

► Check the date the meat was pack-aged. It is recommended that raw meats be cooked within three days. So if it's been in the display cooler for two days, you need to cook it within one day. It is always best to buy the freshest product possible.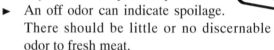

► An off odor can indicate spoilage. There should be little or no discernable odor to fresh meat.

► Raw meat that has spoiled may turn brown, green or rain-bow-colored, develop spots or have a slimy coating and smell "off." Many conditions can contribute to this, such as: unsanitary handling, too long in storage or improper storage temperatures.

► Frozen meats should be thoroughly frozen. If there is mois-ture on the package or frost inside the package, the freezer may not be working properly and the food may have been partly thawed and refrozen.

► Frozen raw meats that have been thawed must be cooked thoroughly before they can be safely refrozen.

Storing

► Meat must be held at refrigeration temperatures of 40°F (4°C) or colder until it is cooked. Most raw meats will keep for about three days under proper refrigeration.

► Some refrigerators have a special "meat keeper" compart-ment which is kept at near-freezing temperature of 32°F (0°C) and may safely lengthen the storage time by two or three days.

▶ Raw meat should not be allowed to contaminate other foods by touching or dripping into them. Store these items on a low shelf, in containers that will not leak onto other foods —like salad ingredients.

▶ It is important to refrigerate meat as soon as possible after buying it. Potentially hazardous foods should be purchased at the end of a shopping trip and stored immediately in the refrigerator or freezer.

▶ Frozen meat should be stored at 0°F (-18°C) or colder in your chest freezer. Note that the freezer on the top of your fridge may not be as cold as a chest freezer, so the stored foods may not maintain the same quality as those stored in a chest freezer.

Water Begins to Freeze
32°F (0°C)

Refrigerator
40°F (4°C)

Frozen Food
0°F (-18°C)

Thawing

▶ Potentially hazardous frozen foods like meat are best thawed in the refrigerator. This may take two days and even longer for large roasts. Safe thawing alternatives include microwave ovens, which can do an excellent job if they have a good defrost setting, or running cool water over the item while it's tightly wrapped in plastic, followed by immediate cooking.

▶ NEVER thaw frozen meats at room temperature. Foods that have been frozen are easy for bacteria to use. They can multiply rapidly on the warmer outer surfaces of thawing foods that are still frozen inside.

▶ Minced or ground raw meats that have been frozen are particularly dangerous because they can contain large numbers of microbes to start with. Thaw them in their own wrapper to prevent further contamination or cross-contamination.

Large cuts
Preparation

NOTE: Any surface or other food that comes in contact with raw meat or the drippings can be contaminated by hitch-hiking or bathing microbes.

▶ To prevent cross-contamination, keep a separate work area or cutting board and tools for handling raw meat. This also makes it easier to be sure you've cleaned all possible contact surfaces when you've finished working.

▶ When preparing **roast beef or lamb** that will be cooked to a **medium-rare** stage, or **pork** that will be cooked to a **medium** stage, don't pierce it deeply with a knife or fork. We can expect pathogens to be on the surface of raw meat and we do not want to inject them deep into the meat where the heat will not be high enough to destroy them.

▶ If a recipe calls for piercing a roast to allow a marinade to penetrate or inserting slivers of garlic, make the cuts shallow and on an angle across, rather than straight down into the meat.

▶ If a roast is going to be cooked to a well done stage, pathogens will be destroyed by the heat; we don't need to be concerned if these foods are pierced during preparation.

▶ After you have finished preparing the raw meat, all work surfaces and tools should be washed and then sanitized using a solution of water and chlorine bleach. The recommended ratio is one ounce (30 ml) of bleach per gallon (4 L) of water.

▶ Don't forget to wash your hands after handling raw meats.

Cooking

A metal probe thermometer is one of the very best investments you will ever make to ensure the health of your family. These inexpensive items can be found in cooking specialty shops or scientific supply houses. They're available in dial or digital types. Since they work on a bimetallic coil principle, there is no danger from mercury poisoning or contamination if you break one of these essential kitchen gadgets. Don't forget to sanitize it before you use it.

NOTE: Insert the thermometer **after** the bacteria on the surface of the roast have been destroyed by the heat of the oven.

▶ It is recommended that **solid roasts of beef or lamb** be cooked to a minimum internal temperature of 145°F (63°C). By the time the center reaches that temperature, the outer surface will be hot enough to kill any bacteria that may be present.

▶ You may decide to cook **solid roasts of pork** from an inspected herd to a medium stage with a minimum internal temperature of 160°F (71°C). If you want more assurance you could cook pork to a well done stage of 170°F (77°C).

▶ It is recommended that roasts of **wild game, or rolled or deboned beef, or stuffed lamb or pork** be cooked to a minimum internal temperature of 170°F (77°C).

Safe Cooking
170°F (77°C)

Boiling Water
212° F (100°C)

Hot Soup
140° F (60°C)

▶ Don't leave the thermometer in the meat while it is cooking. Use it when you are checking for "doneness." The metal probe or glass can pick up the temperature of the air in the oven, and the accuracy of your reading can be affected. Probe thermometers are not designed to reach high oven temperatures and prolonged exposure could blow off the top of your thermometer.

▶ When checking the temperature with a probe thermometer, the tip must not touch bone. Bone has a different density and conducts heat more quickly than the meat and so your reading will be inaccurate.

▶ Partial cooking followed by a finishing stage at a later time is a risky practice. Partially cooked food may feel hot on the surface when we touch it. But, in fact, it means we have warmed the inside of it to the microbes' favorite temperature and softened it making it easier for them to use. The partial cooking and cooling process allows microbes time to grow to dangerous levels.

Steaks, Chops, Ribs and Other Cuts
Preparation

NOTE: Any surface or other food that comes in contact with the raw meat or the drippings can be contaminated by hitch-hiking or bathing microbes.

▶ To prevent cross-contamination, keep a separate work area or cutting board and tools for handling raw meat. This also makes it easier to be sure you've cleaned all possible contact surfaces when you've finished working.

▶ When preparing a cut of **beef or lamb** that will be cooked to a **medium-rare** stage of 145°F (63°C) or **beef, lamb or pork** that will be cooked to a **medium** stage of 160°F (71°C), don't pierce it with a knife or fork. We can expect pathogens to be on the surface of a cut and we do not want to inject them into the meat where the heat will not be high enough to destroy them.

▶ Some recipes advise puncturing the meat with a fork to allow a marinade to penetrate the food. This is not recommended unless the meat will be cooked to a well done stage of 170°F (77°C).

▶ If a cut of meat is going to be cooked to a well done stage, any pathogens will be destroyed by the heat; we don't need to be concerned if these foods are pierced during preparation.

▶ After you have finished preparing the raw meat, all work surfaces and tools should be washed and then sanitized using a solution of water and chlorine bleach. The recommended ratio is one ounce (30 ml) of bleach per gallon (4 L) of water.

▶ Don't forget to wash your hands before doing any other tasks or chores.

Cooking

▶ It is recommended that **beef or lamb** steaks, chops, and ribs be cooked to an minimum internal temperature of 145°F (63°C). By the time the center reaches that temperature, the outer surface will be hot enough to kill any bacteria that may be present.

▶ If you intend to cook **beef or lamb** steaks, chops, or ribs to a **medium-rare or medium** stage, or **pork** to a **medium** stage, use only tongs or a flipper to turn them. A fork may transport harmful bacteria from the surface to the center where the temperature will not be high enough to destroy them.

▶ You may decide to cook cuts of **pork** from an inspected herd to a medium stage with a minimum internal temperature of 160°F (71°C). If you want more assurance you could cook pork to a well done stage of 170°F (77°C).

▶ It is recommended that any cuts of **wild game, or rolled or stuffed beef, lamb or pork** be cooked to an minimum internal temperature of 170°F (77°C). These temperatures will destroy any pathogens that may be in the meat.

▶ Partial cooking followed by a finishing stage at a later time is a risky practice. Partially cooked food may feel hot on the surface when we touch it. But, in fact, it means we have warmed the inside of it to the microbes' favorite temperature and softened it making it easier for them to use. The partial cooking and cooling process allows microbes time to grow to dangerous levels.

Boiling Water
212° F (100°C)

Safe Cooking
170°F (77°C)

Hot Soup
140° F (60°C)

Steaks, Chops and Burgers Touch Test

Use the back of a teaspoon or fork
(Times could change if there is a wind or cool temperatures
and you are cooking out of doors)

MEDIUM - RARE meat gives easily when touched. A hint of juice appears on the surface.
(Minimum four minutes per side for one-inch thick)

MEDIUM meat feels firmer but slightly springy, and juices begin to appear on the surface.
(Minimum Five minutes per side for one-inch thick)

WELL DONE meat is covered with juices. It is firm to the touch and does not yield to pressure.
(Minimum six minutes per side for one-inch thick)

Minced or Ground Meat
Preparation

NOTE: Any surface or food that comes in contact with the raw meat or the drippings can be contaminated by hitch-hiking or bathing microbes.

▶ To prevent cross-contamination, keep a separate work area or cutting board and tools for handling raw meat. This also makes it easier to be sure you've cleaned all possible contact surfaces when you've finished working with the raw food.

▶ If your recipe requires a marinating period, it should always be done in the refrigerator. Some cooks prefer to leave the item marinating for a longer time when it is refrigerated.

▶ After you have finished preparing the raw meat, all work surfaces and tools should be washed and then sanitized using a solution of water and chlorine bleach. The recommended ratio is one ounce (30 ml) of bleach per gallon (4 L) of water.

Cooking

► Ground meat patties should be cooked to a well done temperature of 170°F (77°C) to destroy any *E. coli* bacteria that may be in them. Because patties are too thin to actually measure them with a thermometer, they should be cooked until the juices run clear. Cut them open to see inside: **if they are still pink, they need to be cooked longer—especially if the patties are to be eaten by small children.**

► Meat loaves and pasta stuffed with ground meat must be cooked to an internal temperature of 170°F (77°C) for a few minutes. Meat loaves should be tested with a probe thermometer.

► Partial cooking followed by a finishing stage at a later time is a risky practice. Partially cooked food may feel hot on the surface when we touch it. But, in fact, it means we have warmed the inside of it to the microbes' favorite temperature and softened it making it easier for them to use. The partial cooking and cooling process allows microbes time to grow to dangerous levels.

Cooling, Storage and Reheating Meats

► Potentially hazardous foods like cooked meat must be stored in the refrigerator or freezer as soon as possible. If these foods have been allowed to remain in the danger zone for more than two hours, they should be thrown out.

► Refrigerated leftovers should be used within three days and consumed cold out of the refrigerator or heated to piping hot temperatures. Never just "warm up" leftover meat and then eat it. It could be hazardous to your health.

► Make sure leftovers—including chili, soups or casseroles containing meat—are reheated quickly to 170°F (77°C). This will destroy any microbes that have been introduced since the previous cooking stage.

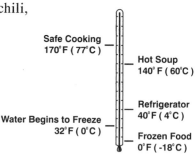

Safe Cooking 170°F (77°C)

Hot Soup 140° F (60°C)

Refrigerator 40°F (4°C)

Water Begins to Freeze 32°F (0°C)

Frozen Food 0°F (-18°C)

► To ensure the heat is evenly distributed throughout the food, stir frequently.

► If you are reheating food for a single or family-size serving, reheat only enough for that occasion and leave the remainder refrigerated. This way the bacteria that may be in the food will not have the added time of reheating and then cooling to grow.

Barbecue Safety

► Keep meats that are to be barbecued at refrigeration temperatures until it is time to grill them.

► If a roast is started in the oven or if ribs are parboiled for finishing on the barbecue, they should be placed on the barbecue immediately after coming from the oven or pot. Don't let them "rest" in the danger zone temperature range.

► If you use a brush to baste raw meat at the beginning of your cooking period, the brush becomes contaminated with the pathogens on the surface of the raw foods. If you baste the cooked, ready-to-eat meat with that same brush, you will recontaminate the cooked food with the pathogens you picked up on the brush. Use a clean brush for cooked food.

► If you intend to cook **beef or lamb** steaks, chops, or ribs to a **medium-rare or medium** stage, or **pork** to a **medium** stage, use only tongs or a flipper to turn them. A fork may transport harmful bacteria from the surface to the center where the temperature will not be high enough to destroy them.

► If raw meats leave drippings on the platter used to carry them to the barbecue, they can recontaminate the foods after they are cooked. Use a clean platter when taking cooked foods off the grill.

Meat Handling — Myths and Facts

Myth: Marinated meat is free of microbes because they are killed by the alcohol in wine or brandy and the neutral pH that bacteria thrive in is altered by adding lemon juice or spices.

Fact: Alcohol is not an efficient sanitizer to begin with. Both the alcohol and the acid in lemon juice are diluted by the juices from the raw meat and so have no real effect on the neutral pH of the food.

Spices like cayenne or mustard taste hot to us, but have no influence on microbes. The only "heat" that will destroy them is thermal heat of 170°F (77°C).

Myth: It is safe to eat foods containing uncooked meat such as steak tartar.

Fact: Bacteria and other pathogens are frequently associated with raw meat. These pathogens are destroyed by cooking meat to safe cooking temperatures.

Only meat from whole muscle cuts of beef or lamb are considered safe when cooked to a medium-rare internal temperature of 145°F (63°C) for a few minutes. And, this is only if they have been prepared without transporting pathogens into the meat with a fork or knife. All other cuts of meat or foods containing meat must be cooked more thoroughly.

Eating foods containing raw meat is not recommended.

Fish and Shellfish

Food from our oceans, rivers and lakes are a significant part of our diet. We have developed many ways of preparing and serving them. Many cultures around the world have favorite and exotic recipes that include fish or shellfish.

These foods are rich in protein, moisture and are neutral on the pH scale. This makes them quite attractive to pathogens—including bacteria, parasites and viruses. Because they also have a fine tissue structure, they are even easier for microbes to use than most meats and poultry.

In highly populated areas, we can expect an increase of microbial contamination to our waters. Viruses and bacteria concentrate in shellfish, especially bivalve mollusks like oysters and clams. They can include *Hepatitis A* viruses and the horrible disease causing *Vibrio vulnificus* bacteria. We can destroy these microbes by thoroughly cooking these mollusks.

Then there are nature's contaminants. For example, algae blooms (red tide) produce the natural toxins which cause Paralytic Shellfish Poisoning. Some fish, like "puffer fish," can become highly toxic by their nature.

As well, some warm-water fish like the scombroid types (i.e. mackerel, tuna, marlin, mahi-mahi and bluefish) can produce histamines, which can cause people to become ill unless the catch is first handled properly by being quickly iced or frozen at sea.

Unfortunately toxins and histamines are not destroyed during cooking.

Marine fish are often hosts to parasites—including *Anisakis* and other round worms and their eggs. These can be a concern to humans if they are ingested in their live state. Freshwater fish or fish that spend part of their life in fresh water may carry tapeworm larvae. However, finding a parasite in fish is considered a natural occurrence and not a form of contamination. Safeguards such as thorough cooking or controlled freezing will destroy these parasites.

All of these factors mean that we must take particular precautions when purchasing, harvesting, handling and preparing fish and shellfish.

Eating Raw Fish or Shellfish

Recipes from many cultures around the world include raw or lightly cooked seafoods. The documented increase in reported cases of foodborne illness from consuming raw fish and shellfish has brought strong warnings from food specialists.

Eating raw seafoods such as oysters, clams, scallops, shellfish and many types of finfish is not recommended.

Raw fish used in sushi or ceviché must be frozen for at least seven days at 0°F (-18°C) to destroy worms and their eggs. Freezing does not destroy bacteria or some parasites. The marinating of fish for foods such as ceviché is not sufficient to destroy pathogens.

Eating raw mollusks is not recommended

Purchasing

When purchasing ready-to-eat seafood like lox or cooked shrimp meat make sure the vendor's display discourages accidental contamination. For example, a seafood display might allow drippings from raw scallops to contaminate surrounding finished seafoods when the scallops are ladled across finished food.

Another common source of seafood contamination includes nestling ready-to-eat and raw products in the same display bed of ice. If you observe these examples during your seafood shopping, bring it to the manager's attention and shop elsewhere where better food handling is practiced. You will reduce your family's risk of encountering foodborne illness.

► Always buy seafoods from a reputable source.
► Check the temperature in the display case. Fish and shellfish should always be held and displayed well below 40°F (4°C) or colder. If the cooler is not cold enough, there could already be countless microbes growing in the food.
► Check the date that the seafood was packaged. It is recommended that seafoods be cooked or frozen within two days of processing. So, if it's been in the display cooler for one day, you need to cook it within one day.

► It is always best to buy the freshest product possible.

► Frozen seafood should be thoroughly frozen. If there is moisture on the package or if there is frost inside the package, the freezer may not be working properly.

► Raw seafood that has been frozen must be cooked thoroughly before it can be safely refrozen.

Finfish

► When buying fish or using your own catch you can look for signs of quality and freshness. The eyes should appear bright, clear—almost alive; the gills reddish; the skin moist with shiny, tightly adhered scales. Fresh fillets have a bright, shiny color without browning.

► Fresh fish shouldn't smell "fishy." A fresh fish aroma is never offensive.

► The flesh of fresh fish should be firm, give slightly when gently pressed, and then spring back into shape. If you are not sure about the freshness of a fish, don't buy it or use it.

► Smoked fish requires refrigeration and should be used within seven to ten days of processing. There should be no signs of deterioration or spoilage including discoloration, mold, sliminess or an unpleasant odor.

► Some fish jerky has been processed in a way that can make it possible to hold it safely at cool room temperatures through salting and thorough drying. The manufacturer's label will indicate whether or not it requires refrigeration.

► When any ready-to-eat fish is sold in a package that has had the air removed, such as in vacuum packaging, it must be in a frozen solid state. *Botulinum type E* bacteria are common in the marine environment. They share the toxin producing abilities of the other strains of *Botulinum,* but they may also reproduce at refrigeration temperatures.

► *Listeria* bacteria can reproduce at cold temperatures as well, and they are another concern for seafood products.

▶ When these foods are frozen, these bacteria, when present, cannot reproduce. If you are thawing frozen fish that has been stored in an airtight package, open it to allow oxygen to reach the fish to prevent the *Botulinum* bacteria from reproducing.

▶ The manufacturer of vacuum packaged, ready-to-eat fish can also make these products "shelf stable" by specialized retorting or other processes. Again, we must read labels carefully.

Shellfish

▶ **Oysters, clams and mussels** in the shell should be alive. They should be tightly closed, or they should close when tapped lightly. Discard any with broken shells. If you are buying them already opened or shucked, they should be plump and the liquor should be clear, not cloudy, with no sour or unpleasant odor.

▶ If you have harvested them yourself, and the shells were broken in the process, they should be used immediately.

▶ Live **crab, lobster and crayfish** are best kept in cold, salt water using five tablespoons of sea salt (120 ml) per gallon (4 L) water with air circulation. Crustacean shellfish will go into shock if they are held in fresh water. Once they have been removed from the water, they should be cooked immediately while they are still alive.

▶ Freshly cooked shellfish meat should have no ammonia-like odor.

▶ Cooked **shrimp and prawns** should be dry and firm.

Storing

▶ Fish and shellfish must be held at refrigeration temperatures of 40°F (4°C), or preferably colder, or frozen until it is cooked.

▶ Seafoods stored in the refrigerator should be used within one to two days.

▶ Live bivalve mollusks are best stored in well ventilated containers covered with a moist paper towel.

▶ Frozen seafoods must be held at 0°F (-18°C) or colder.

▶ Raw fish and shellfish should not be allowed to contaminate other foods by touching or dripping into them. Store the raw seafood on a low shelf, in a container that will not leak into ready-to-eat foods.

▶ It is important to refrigerate seafoods as soon as possible after buying them. It is recommended that the more potentially hazardous foods be purchased at the end of a shopping trip and placed in cold storage immediately.

Water Begins to Freeze 32°F (0°C)

_ Refrigerator
40°F (4°C)

_ Frozen Food
0°F (-18°C)

Thawing

▶ Potentially hazardous foods like frozen fish and shellfish are best thawed in the refrigerator. This may take two days for large fish like whole salmon. Safe thawing alternatives include microwave ovens, which can do an excellent job if they have a good defrost setting, or running cool water over the item wrapped in plastic.

▶ Never thaw frozen seafood at room temperature. Foods that have been frozen are a treat for bacteria, and they can multiply rapidly on the warmer outer surface of foods that are still frozen inside.

▶ If you are thawing frozen fish that has been stored in an airtight package, open it to allow oxygen to reach the fish to prevent the *Botulinum* bacteria from reproducing.

▶ Most fish or shellfish can be cooked directly from the frozen state.

Preparation

NOTE: Any surface that comes in contact with the raw seafood or the drippings can be contaminated by hitch-hiking or bathing microbes.

▶ To prevent cross-contamination, keep a separate work area or cutting boards and tools for preparing or handling raw seafoods. This also makes it easier to clean all possible contact surfaces when you've finished working.

▶ After you have finished preparing the raw fish or shellfish, all work surfaces and tools should be washed and then sanitized using a solution of water and chlorine bleach. The recommended ratio is 1 ounce (30 ml) of bleach per gallon (4 L) of water.

▶ If your recipe requires a marinating period, it should always be done in the refrigerator. Some cooks prefer to leave the item marinating for a longer time when it is refrigerated.

▶ Fish that will be consumed raw in sushi or ceviché recipes, must be deep frozen for a minimum of seven days at 0°F (-18°C) to destroy round worms and their eggs.

▶ Shellfish, including oysters, clams, mussels, scallops, shrimp, prawns, crab, crayfish and lobster should be cooked before they are eaten. These seafoods often carry pathogens that are not destroyed by freezing.

Cooking

▶ Properly cooked fish will flake easily with a fork and should be opaque and firm. Allow 10 minutes of cooking time per inch of thickness and turn halfway through the cooking time. Add five minutes to the total cooking time if the fish is wrapped in foil or cooked in a sauce.

▶ Shellfish, including clams, oysters, mussels, scallops, shrimp, prawns, crab, crayfish and lobster should be cooked to 170°F (77°C).

Boiling Water
212° F (100°C)

Safe Cooking
170°F (77°C)

Hot Soup
140° F (60°C)

► Continue boiling for three to five minutes after the shells of clams, oysters, mussels or scallops open.

► Light steaming of mollusks is not recommended so you should continue to steam cook them for four to nine minutes after the shells open.

► Discard any mollusks if their shells do not open during cooking.

► Shucked oysters should be boiled for at least three minutes, pan fried for at least 10 minutes or baked thoroughly.

► Crab, crayfish and lobster should be boiled for 20 minutes to be sure they have reached temperatures high enough to destroy any pathogens.

► Shrimp and prawns should be cooked thoroughly using whatever method suits your recipe.

► Partial cooking followed by a finishing stage at a later time is a risky practice. Partially cooked food may feel hot on the surface when we touch it. But, in fact, partial cooking means we have warmed the inside of it to the microbes' favorite temperature, softened it and made it easier for them to use.

Cooling, Storage and Reheating

► Potentially hazardous foods like cooked seafoods must be stored in the refrigerator or freezer as soon as possible. If these foods have been allowed to remain in the danger zone for more than two hours, they should be thrown out.

► Refrigerated leftovers should be used within two days, at most, and consumed cold out of the refrigerator or heated to piping hot temperatures. Never just "warm up" leftover seafood and then eat it. It could be hazardous to your health.

► To destroy any microbes that may have been introduced since the previous cooking stage, make sure that leftovers are reheated quickly to 170°F (77°C).

► If you are reheating fish or shellfish for a single or family-size serving, reheat only enough for that occasion and leave the rest refrigerated. This way any bacteria still in the remaining food will not have additional time to grow during reheating and cooling.

Stuffed Fish
▶ Because stuffing contains many ingredients that are handled a lot, there can be many pathogens in the food. If the uncooked stuffing contains raw seafood, we can expect microbes to be in it. Stuffing should be cooked to 170°F (77°C).

Barbecue Safety
▶ Keep seafoods that are to be barbecued at refrigeration temperatures until it is time to grill them.
▶ If you want to baste the seafood with leftover marinade, the marinade should first be brought to a boil to destroy any pathogens that may have been in the raw food. It is important to protect cooked food from contamination by raw food.
▶ If you use a brush to baste raw seafoods at the beginning of your cooking period, the brush becomes contaminated with the pathogens on the surface of the raw foods. If you baste the cooked, ready-to-eat seafood with that same brush, you will recontaminate the cooked food with the pathogens you picked up on the brush. Use a clean brush for cooked seafood.
▶ The cooked seafood can become contaminated by drippings on the platter used to carry raw foods to the barbecue. Use a clean platter when taking cooked foods off of the grill.

Seafood Handling — Myths and Facts
Myth: Marinated seafoods, like the fish in ceviché, are free of microbes because they are killed by alcohol in wine or brandy, and the neutral pH that bacteria thrive in is altered by adding lemon juice or spices.

Fact: Alcohol is not an efficient sanitizer to begin with. Both the alcohol and the acid in lemon juice are diluted by the juices from the raw seafood and so have no real effect on the neutral pH of the food.

Spices like cayenne or curry taste hot to us, but have no influence on microbes. The only "heat" that will destroy them is thermal heat of 170°F (77°C).

Myth: It is safe to eat raw oysters and other mollusks.

Fact: Bivalve mollusks can become contaminated with pathogenic viruses such as *Hepatitis A* or bacteria such as *Vibrio vulnificus*. This contamination comes from the waters where the bivalves are harvested and can be especially heavy where there are many people in the area. Eating raw oysters, clams, mussels or scallops is not recommended.

Myth: You can tell if clams, oysters or mussels are contaminated with dangerous toxins from algae bloom (red tide) by rubbing them on your lips. If they make your lips tingle, they are contaminated; if they don't, they are safe to eat.

Fact: Paralytic Shellfish Poisoning can cause such serious illness that biological contamination levels are monitored through tests conducted by government authorities. Areas where contamination is found are closed to harvesting until they become safe again.

Myth: It is safe to eat the raw fish in sushi or ceviché as long as the fish you use is fresh and has been refrigerated.

Fact: Raw fish used in sushi or ceviché must be frozen for at least seven days at 0°F (-18°C) to destroy worms and their eggs. Freezing does not destroy bacteria or some parasites.

Casseroles

Casserole cooking is popular with many cultures. Recipes for delightful combinations of foods come to us from almost every region in the world. Tasty, oven-baked dishes, like lasagna, shepard's pie or baked beans are the feature of hearty meals for families everywhere.

This versatile cooking method combines any number of ingredients and spices that are mixed, layered or combined in a pan or dish, and baked in the oven. The resulting blend of flavors and textures can be quite delicious as well as nutritious.

We combine different types of foods, and if some of them are potentially hazardous, they may contain pathogens to begin with. When we prepare these foods, we often handle the ingredients, and so, they can become contaminated with pathogens from our hands.

When we cook these foods, we can provide a soft, nutritious, moist, warm environment for pathogens if we give them the opportunity to grow in the food.

This combination of a variety of ingredients which are handled a lot means we must pay close attention to the principles of correct handling and temperature control.

Preparation

NOTE: Remember that any surface coming in contact with raw foods or meat drippings can be contaminated with the microbes that are hitch-hiking on or in the food.

▶ To prevent cross-contamination, keep a separate work area or cutting boards and tools for preparing or handling raw foods. This also makes it easier to clean all possible contact surfaces when you've finished working.

▶ After you have finished preparing the raw foods, all work surfaces and tools should be washed and then sanitized using a solution of water and chlorine bleach. The recommended ratio is one ounce (30 ml) of bleach per gallon (4 L) of water.

▶ Casseroles often contain many potentially hazardous foods. They are also handled a great deal during the preparation stage. Some ingredients may be held at room temperature while other ingredients are being prepared. So, their safety margin can be shorter than that of foods that have just come from the refrigerator.

▶ It is recommended that preparation be organized in such a way that the potentially hazardous foods are exposed to danger zone temperatures for as short a time as possible.

Cooking

▶ Casseroles must be cooked to an internal temperature of 170°F (77°C) for several minutes to kill any microbes in the food.

▶ When checking the temperature with a probe thermometer, the reading should be taken from the center of the food without contacting the sides or bottom of the dish or bones in the food.

▶ Partial cooking followed by a finishing stage at a later time is a risky practice. Partially cooked food may feel hot on the surface when we touch it. But, in fact, we have warmed the inside of it to the microbes' favorite temperature and softened it, making it easier for them to use. The partial cooking and cooling process will allow time for them to grow to dangerous levels.

Cooling, Storage and Reheating

▶ Potentially hazardous foods like casseroles must be stored in the refrigerator or freezer as soon as possible. If these foods have been allowed to remain in the danger zone for more than two hours, they should be thrown out.

▶ Refrigerated leftovers should be used within two days at most and consumed cold out of the refrigerator or heated to piping hot temperatures. Never just "warm up" the leftover casserole and then eat it. It could be hazardous to your health.

► Make sure these foods are reheated quickly to 170°F (77°C). That will destroy any microbes that have been introduced since the previous cooking stage.

► If you are reheating a casserole for a single or family-size serving, reheat only enough for that occasion and leave the rest refrigerated. This ensures that any bacteria that may be in the remaining food will not have added time to grow during the reheating and cooling periods.

► If you don't expect to be able to use all of the leftover casserole within two days, wrap and store what you won't use in the freezer.

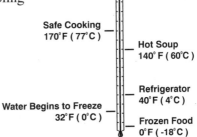

Safe Cooking
170°F (77°C)

Hot Soup
140° F (60°C)

Refrigerator
40°F (4°C)

Water Begins to Freeze
32°F (0°C)

Frozen Food
0°F (-18°C)

Soups, Stews and Gravies

These versatile and nourishing foods have been an important part of the diet of many cultures around the world; probably since the time the first pot was invented.

Some cooks like to make large batches of soup-base or broth and refrigerate or freeze it. Then, when soup is on the menu, fresh ingredients are added to the reheated broth and a delicious dish is ready in a short time.

In the past, the traditional stew or soup "pot" sat on the back of the stove, and was used to collect and stew leftover meats, vegetables and cooking water. The pot simmered away and over the course of several days the resulting brew changed depending on the daily ingredients added. A "new" soup was always in the pot. This is now considered to be a risky practice and is NOT recommended.

Crock-pots or slow-cookers have become popular with many modern cooks. Ingredients are placed in the pot early in the morning and the mixture is cooked at a low heat throughout the day. At dinner time, a full meal is ready with no further preparation or waiting. This can be done safely, however, many cooks are not aware of the potential hazards involved.

Some gravies are made by reducing a broth containing seasoning vegetables or herbs, over the course of several hours until the final mixture is a thick, concentrated, sauce.

Proper preparation and storage of these foods is important because microbes often find the perfect environment in soups, stews and gravies. If the cook does not bring the broth back to a slow boil after each new batch of ingredients is added, the temperature, nutrients, and moisture conditions exist for pathogens that may be in the food to multiply rapidly.

Soups, stews and gravies may include several potentially hazardous ingredients so cooking temperatures require close attention. Cooling and storage of large batches can be troublesome if there is limited refrigeration space or a limited supply of storage containers.

Preparation

NOTE: Remember that any surface that comes in contact with raw foods or meat drippings can be contaminated by hitch-hiking or bathing microbes.

▶ To prevent cross-contamination, keep a separate work area or cutting boards and tools for preparing or handling raw foods. This also makes it easier to clean all possible contact surfaces when you've finished working.

▶ After you have finished preparing the raw foods, all work surfaces and tools should be washed and then sanitized using a solution of water and chlorine bleach. The recommended ratio is one ounce (30 ml) of bleach per gallon (4 L) of water.

▶ Soups stews and gravies often contain many potentially hazardous ingredients. They are also handled a great deal during the preparation stage. Some ingredients may be held at room temperature while other ingredients are being prepared. Foods that have been held in the danger zone temperature range have a shorter safety margin than foods that have just come from the refrigerator.

▶ It is recommended that preparation be organized in such a way that the hazardous foods are exposed to danger zone temperatures for as short a time as possible.

▶ Dried legumes and whole grains require a thorough rinse in cold water to remove dirt and debris that may contain pathogens.

▶ Many legumes require a soaking period, usually overnight, to tenderize them and reduce the cooking time. Because they are in their raw state, these foods are safe to soak in cold water at room temperature for up to 12 hours.

Cooking

► Most soups have many ingredients added throughout the cooking process. It is very important to bring these foods back to an internal temperature of at least 170°F (77°C) for several minutes each time you add new ingredients.

► But, by bringing soups, stews and gravies back to a boil 212°F (110°C) each time you add new ingredients (including vegetables, meats or fresh spices) you can be sure you've destroyed any pathogens that may be carried on them.

Safe Cooking
170°F (77°C)

_ Boiling Water
212° F (100°C)

_ Hot Soup
140° F (60°C)

► When checking the temperature with a probe thermometer, the reading should be taken from the center without contacting the sides or bottom of the pot.

► Most dried beans, peas, whole grains and lentils take a long time to cook. They need to be brought to a boil and then simmered for several hours. The temperature should stay above 140°F (60°C) if the food will be simmered for more than one hour.

► To be sure all parts of the soup, stew or gravy stay hot enough to prevent bacteria from growing, stir simmering foods often.

► By stirring frequently you will distribute oxygen as well as heat through the food. The oxygen and heat will prevent growth of any *Botulinum* spores that may have been hitch-hiking on vegetables like onions or garlic.

NOTE: Temperatures over 170°F (77°C) will kill pathogens that have multiplied, but the toxins that they have been produced may not be destroyed by heat.

You must prevent the bacteria from reproducing in the first place. If you keep these foods hot while cooking and cool and refrigerate them quickly, the bacteria cannot multiply rapidly.

Cooling, Storage and Reheating

► Once the ingredients in soups, stews and gravies have been cooked they are soft, moist and nutritious so any bacteria that are introduced now have the conditions they need to grow. Any spores that are on the dried or raw foods can now vegetate and grow.

► Legumes, grains and vegetables become hazardous after cooking and must be stored in the refrigerator or freezer as soon as possible after cooking.

► Some soups and stews made with legumes can be very dense, for example split pea soup or chili. In order to cool them quickly enough to keep microbes from growing in them, it may be necessary to use a shallow pan and stir the food as it cools.

► Metal dishes work best because the metal will help draw heat from the soup. By using a shallow dish you create more surface area for the heat to escape from. Stirring also helps the heat escape more quickly.

► You can speed up this process by placing the dish on a bed of ice or setting it in cold water. Cooling wands (food grade plastic, gel filled, wands that are stored in the freezer until needed) can be used to stir the soup or stew.

► Don't place hot foods into plastic dishes. Some plastics may not be designed to hold hot foods. Also, they don't transfer heat as efficiently as metal equipment.

► If soups, stews or gravies have been allowed to remain in the danger zone for more than two hours, they should be thrown out.

► Refrigerated leftovers should be used within two days and consumed cold out of the refrigerator or heated to piping hot temperatures. Never just "warm up" leftover soup, stew or gravy and then eat it.

► Make sure that these foods are reheated quickly to 170°F (77°C). This will destroy any microbes introduced since the previous cooking stage.

Safe Cooking
170°F (77°C)

Hot Soup
140° F (60°C)

Refrigerator
40°F (4°C)

Water Begins to Freeze
32°F (0°C)

Frozen Food
0°F (-18°C)

► If you used leftovers in your soup or stew, keep in mind that the safety margin of the entire finished product is only as long as the oldest ingredient. So, if your leftover beef had already been stored in your refrigerator for two days, you will need to use the entire soup or stew within one day.

► If you are reheating food for a single or family-size serving, reheat only enough for that occasion and leave the remainder refrigerated. If there's any bacteria in the food, they won't have additional time to grow during reheating and cooling.

► If you don't expect to be able to use all of the leftover soup, stew or gravy within two days, store the remainder in the freezer.

► Leftover soups, stews and gravies should be brought to a boil before they are served. They should be frozen or reheated only once.

Crock-Pot or Slow-Cooking

Follow correct procedures when slow-cooking soups and stews, to ensure they're safe. If the temperature of the food is allowed to move in and out of the danger zone range, slow-cooking creates a perfect environment for bacteria to grow quickly.

If you begin with cold or cool ingredients, it could easily take more than two hours for the contents of the cooker to reach a temperature high enough to destroy bacteria. Even if the food is boiled before it is served it may not be safe because certain bacteria produce pathogenic toxins which may not be destroyed by heat.

Preparation

► Choose a good quality slow-cooker with a dependable thermostat. Use a probe thermometer to test the actual temperature of the food against the setting of your slow cooker, and adjust for any difference or discrepancy.

► Use boiling water or broth when you first fill the cooker, to raise the temperature as high as possible from the start.

► Make sure the temperature of the soup or stew is above 170°F (77°C) before turning the thermostat down to the slow-cooking temperature. Test with a probe thermometer.

Take the reading from the center without contacting the sides or bottom of the pot.

► Set the slow-cooker temperature to hold the food above 140°F (60°C) to make sure that any bacteria introduced into the food cannot grow quickly enough to cause illness.

Soups, Stews and Gravies — Myths and Facts

Myth: Vegetarian soups do not need to be kept hot or be refrigerated quickly. Because they don't contain animal or dairy products, nothing dangerous can grow in them.

Fact: Once vegetables, beans, peas or lentils are cooked, they become potentially hazardous foods. They provide the right environment for bacteria to grow after they are cooked. These foods must be held at the correct temperatures to slow the growth of any dangerous bacteria that may be in them. Microbes will grow in any food that provides the conditions they need.

Myth: Foods like chili and curry gravies are safe because they have strong spices in them.

Fact: Spices like cayenne and curry taste hot to us, but they have no influence on microbes. The only "heat" that will affect the microbes is thermal heat of 170°F (77°C).

Myth: Soups, stews and gravies only taste better and better as they age each day. Time gives the flavors a chance to blend and develop.

Fact: If these foods are held at danger zone temperatures for more than two hours (including cooling and reheating time), enough dangerous bacteria can grow in them to make someone ill after eating them.

Refrigeration slows the growth of bacteria but does not stop it. Soups, stews and gravies can safely be held at refrigeration temperatures of 40°F (4°C) for up to two days. Flavors continue to blend in foods held in the refrigerator or freezer. Bacteria cannot grow in frozen foods, making this a safer method for blending flavors.

Dairy Foods

Dairy products like milk are considered to be wholesome and nourishing foods. So, they are often a large part of the diet of infants, elderly and ill people. These people are also more susceptible to food poisoning.

Dairy products provide the rich, moist, neutral pH environment that microbes need to reproduce and survive. If these foods are not stored at refrigeration temperatures of 40°F (4°C) or colder, any microbes in the foods will have the right conditions to grow.

Ice cream and soft cheeses have been associated with outbreaks of Listeriosis and Salmonellosis. Other dangerous microbes can also be spread by dairy products. Because these foods provide such a good environment for microbes, **unpasteurized dairy products** are considered extremely **dangerous** to our health.

Dairy producers must follow strict controls to ensure their products are not contaminated. These products are labeled with a "best before" date because some spoilage microbes continue to grow under refrigeration temperatures (the colder the temperature, the more slowly they grow). Given enough time, they can grow to levels that can spoil the food items. For example, each time you leave a container of milk on the kitchen counter any spoilage microbes that may be in it will reproduce more quickly—and that would affect the validity of the "best before" date.

Curdling is a visible sign the milk has spoiled. But, the microbes that can make us sick can be present in high numbers with **no visible sign** that the milk is contaminated.

If mold is present on the surface of a hard cheese like cheddar and mozzarella, it is recommended that at least 1 inch (2.5 cm) be cut off. Remember that the fuzzy things you see are like flowers and the stems and root systems are under the surface and can't be seen with the naked eye. Some molds produce mycotoxins as they grow. The toxins are not be destroyed by cooking. Mycotoxins are thought to cause cancer.

Mold & its spores

In fact, it is recommended to not open the package once mold is visible. Not only can the spores contaminate your kitchen, our bodies can provide the environment for pathogenic molds to grow. Spores can germinate on the surface of the eye causing blindness, or in the lungs causing respiratory problems.

Soft dairy products such as sour cream, should be thrown out if they are contaminated with mold.

Purchasing

► Always buy dairy products from a reputable source.

► Unpasteurized milk or products made from unpasteurized milk are not recommended. Choose only pasteurized dairy products, especially if they will be consumed by someone who is very young, elderly or ill. These people are at at high risk to food poisoning.

► Check the temperature in the display case. Dairy products should always be held below 40°F (4°C). If the cooler is not cold enough there could be an unacceptable number of microbes in the food.

► Check the "best before" date on the pack- age. It is always best to buy the freshest product possible.

► Tiny white spots on the surface of hard cheeses like cheddar and mozzarella can indicate the early stages of mold growth. The salts in some cheeses will come to the surface. This is another sign that the cheese is spoiling after being improperly stored for too long.

► Liquid inside the unbroken packaging of hard cheeses can indicate that the product has been frozen, or has been allowed to "sweat" by being held at warm temperatures. This added moisture can cause the product to spoil more quickly.

► Any foods that have been frozen are easier for microbes to use. After they thaw, they will spoil more quickly than fresh foods. Foods that contain dairy products and have been frozen should be used as quickly as possible. You can keep them up to two days in the refrigerator. Hard cheeses will keep for several days.

► Some cheese is packaged in cans. These foods are partially processed and cannot be handled in the same way we handle other "canned" foods. If the label reads "Keep Refrigerated," these products must be held at refrigeration temperature at all times. These partially processed foods have a shelf life, stored unopened in the refrigerator, of six months from the time they are packaged. Look for the "best before" date. If they are not marked, use them quickly because that six month shelf life includes shipping and display time.

Water Begins to Freeze
32°F (0°C)

Refrigerator
40°F (4°C)

Frozen Food
0°F (-18°C)

Storing

► Dairy products must be refrigerated at temperatures of 40°F (4°C) or colder.

► They should be purchased at the end of a shopping trip and refrigerated as soon as possible.

► Unsalted butter should be stored in the refrigerator because some bacteria can survive and reproduce on or in it.

► Keep in mind that each time you leave a dairy food—for example, milk—at room temperature, any microbes in the food will grow rapidly. If these foods are held in the danger zone temperature range for more than a total of two hours, they should be thrown out.

Preparation

NOTE: Many dairy products are ready-to-eat foods. Care should be taken to prevent cross-contamination from raw foods or work surfaces and utensils. To prevent cross-contamination, keep a separate work area or cutting board and tools for ready-to-eat foods.

Cooking or Mixing

Dairy products are often served without ever cooking them at a high temperature. Because there is no added "kill temperature" in the process, we need to pay close attention to how we store, handle and prepare them.

► When preparing foods that contain dairy products, you can plan your steps in a way that will keep these foods at refrigeration temperatures of 40°F (4°C) as much as possible.

► For example, if your recipe includes sour cream which is mixed with other ingredients at the end of the process, you should leave the sour cream in the fridge until the moment it is needed. If you leave it on the counter for 45 minutes, both the remaining sour cream and your creation will have only 1 and 1/4 hours left to safely remain at room temperature.

Dairy Sauces & Custards

If left unrefrigerated, or below the safe hot holding temperature of 140°F (60°C), sauces made with dairy products provide an excellent environment for harmful bacteria to grow in. Sauces and custards that are prepared in advance or left over must be cooled quickly to slow the growth of harmful bacteria.

Some custard recipes call for raw egg whites to be mixed into the base ingredients which have been cooked. Since the raw eggs may carry harmful bacteria, they can contaminate the custard.

It is safer to use a recipe that requires further cooking of the egg whites or to use pasteurized egg products as an alternative.

Cooling, Storage and Reheating

► To cool a sauce or custard rapidly, place it in a shallow dish and stir it as it cools.

► Metal dishes work best because the metal will help draw heat from the sauce. By using a shallow dish, you create more surface area for the heat to escape from. Stirring helps the heat escape more quickly.

► You can speed up this process by placing the dish on a bed of ice or setting it in cold water. Cooling wands (food grade plastic, gel-filled wands that are stored in the freezer until needed) can be used to stir the sauce or custard.

► Placing hot foods into plastic dishes is NOT recommended. Some plastics may not be designed to hold hot foods and are less efficient than metal for transferring heat.

► Refrigerated leftovers should be used within two days and consumed cold out of the refrigerator or heated to piping hot temperatures. Never just "warm up" leftover dairy sauces or custards and then eat them. It could be hazardous to your health.

► Make sure that these foods are reheated quickly to 170°F (77°C). This will destroy any microbes introduced since the previous cooking stage.

► If you are reheating food for a single or family-size serving, reheat only enough for that occasion and leave the remainder refrigerated. If there's any bacteria in the food, they won't have additional time to grow during reheating and cooling.

Safe Cooking 170°F (77°C)

Hot Soup 140°F (60°C)

Refrigerator 40°F (4°C)

Water Begins to Freeze 32°F (0°C)

Frozen Food 0°F (-18°C)

Dips

Many dips contain dairy products such as sour cream, cream cheese or mayonnaise. If they are purchased pre-made, they need to be handled and stored according to the manufacturer's directions.

When preparing home-made dips, it is best to add the dairy products in the later stage of the recipe keeping the ingredients refrigerated as long as possible.

Dips should be used within two hours from the time they were removed from the refrigerator. Any remaining dip that is not used in that time should be thrown out.

Dairy Substitutes

Because they are made from vegetable (and other edible oils) we may not think of dairy substitutes, including soy products, as perishable.

Dairy substitutes should be handled as a potentially hazardous food because they give pathogenic microbes the environment they need to survive. These foods must be refrigerated or frozen. Always read the label and follow the manufacturer's instructions for storing dairy substitutes.

Dairy Products — Myths and Facts

Myth: Foods like dips are free of microbes because the neutral pH that bacteria thrive in, is altered by lemon juice or spices.

Fact: The acid in lemon juice will be diluted by the moisture in the dairy products and will have no real effect on the neutral pH of the food.

Spices like cayenne taste hot to us, but they have no influence on microbes. The only "heat" that will affect the microbes is thermal heat of 170°F (77°C).

Myth: Milk curdles when it spoils so you can use it until it goes lumpy or smells sour. Mold is the only other thing that shows us that dairy products are not good to eat or drink.

Fact: Milk and other dairy products are an excellent environment for many pathogens to survive or grow in. Unfortunately most of them cannot be seen without a microscope. So, we must depend on correct handling procedures by the supplier, the distributor, and finally, ourselves.

Essentially, the correct procedures amount to following correct time and temperature rules. Keep hot foods hot and cold foods cold and throw out any dairy products that have been in the danger zone temperature range for more than two hours.

Eggs

Eggs have long been considered one of humankind's perfect foods. They are versatile, nourishing and plentiful. They can be prepared in a wide variety of ways and are used in many recipes.

Because there is such a high demand for this food, it is widely distributed in most countries throughout the world. When eggs are distributed, any harmful bacteria that may be on or in them travels along with them.

When an egg is laid it could become contaminated with waste material from the hen. If the hen is carrying *Salmonella* bacteria, it can be passed to the eggshell. Some commercial egg producers wash the eggs to remove the waste material, but bacteria can still be present.

Salmonella enteritidis is a strain of pathogenic bacteria that can be found in the ova of a hen. This means some eggs can be formed with the bacteria inside the shell even before they are laid. Reports of outbreaks associated with this strain of bacteria are becoming more common around the world. The number of poultry affected by *Salmonella enteritidis* varies from region to region and country to country.

Remember that the *Salmonella* bacteria commonly found in chicken does not normally make the chicken sick. These bacteria seldom cause any harm to the bird. People are the creatures that get sick from consuming this type of bacteria.

Because eggs are such a nourishing food, they are a popular part of the diet for very young, elderly or ill people who are most susceptible to food poisoning. Eggs prepared for people in this high-risk group should be cooked thoroughly.

Also, because eggs are so widely distributed, there may be no way for the average person to know where an egg came from. *Salmonella* bacteria can travel and survive anywhere when their growth conditions are present. Some countries control the extent of geographic trade in eggs.

Eating raw eggs in any form is no longer recommended. Use pasteurized egg products instead.

Eating raw eggs is not recommended

Purchasing

▶ Always buy eggs from a reputable source.

▶ Check the temperature in the display case. Eggs should always be held and displayed below 40°F (4°C). If the cooler is not cold enough there could already be countless bacteria growing on or in them.

▶ Check the "best before" date on the carton. It is always best to buy the freshest product possible.

▶ Don't buy cracked eggs. If there are *Salmonella* or other organisms on the shell, they can easily enter the egg.

▶ It is best to buy pasteurized egg products for use in recipes, such as Caesar salad dressing or egg nog, that will not be cooked.

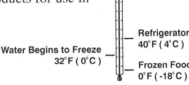

Water Begins to Freeze
32°F (0°C)

Refrigerator
40°F (4°C)

Frozen Food
0°F (-18°C)

Storing

▶ To control the growth of any *Salmonella* bacteria that may be on or in them, eggs must always be stored at refrigeration temperatures of 40°F (4°C) or colder.

▶ Eggs should be stored in their original carton to protect them from accidental damage and moisture. The door of the refrigerator is usually warmer, so it is best to put them on a shelf inside the refrigerator.

▶ It's not necessary to wash commercially-produced eggs before storing them. Washing removes the protective oil that is found naturally on eggs so producers often replace it with an edible mineral oil. The "best before" date on the carton of commercially sold eggs will indicate how long you can safely use them.

▶ Raw eggs taken directly from the nest should be cleaned, refrigerated immediately, and used within three weeks. Only freshly laid eggs should be collected.

► Wash the fresh eggs as soon as possible in hot water at 110°F (44°C) using a soft bristled brush to prevent the fecal matter and wastes on the surface from drying hard. It is important to be very gentle because the shells are fragile and bacteria could be forced into the pores. Rinse them in warm —not cold—water and place them in a sanitizing solution of one ounce (30 ml) of bleach per gallon (4 L) of warm water for a few minutes. Allow them to air dry before storing them in the refrigerator.

► Only very fresh, uncracked, eggs should be frozen. Eggs that have been frozen are recommended only for use in foods such as cakes and fancy breads that are baked at a high temperature for a long time.

► Raw eggs that have been frozen must be cooked thoroughly before they can be safely refrozen.

► Pasteurized egg products are perishable and must always be held at refrigeration temperatures or frozen.

Preparation

NOTE: Any surface that comes in contact with raw eggs or eggshells can be contaminated with the bacteria that are hitch-hiking on or in them.

► To prevent cross-contamination with other foods, use a separate work area or cutting boards and tools for preparing or handling raw eggs. This also makes it easier to clean all possible contact surfaces when you're finished.

► After you have finished preparing the raw eggs, all work surfaces, dishes and tools should be washed and then sanitized using a solution of water and chlorine bleach. The recommended ratio is one ounce (30 ml) of bleach per gallon (4 L) of water.

► Cook eggs thoroughly for people who are very young, elderly or ill because they are vulnerable to food poisoning. It takes approximately four minutes to destroy *Salmonella* in a fried egg and seven minutes to ensure that a boiled egg is safe for people at risk. The yoke and white should be firm, not runny. Scrambled eggs should be cooked dry.

► Do not taste raw cake batter, cookie batter, or any other food that contains uncooked eggs. They could contain *Salmonella* bacteria.

► Some recipes call for raw whipped egg whites to be added to the dish. Eating raw eggs in any form is not recommended. Use pasteurized egg products instead.

► In some recipes, such as Caesar salad dressing, it is possible to substitute commercial, pasteurized mayonnaise (base ingredients are eggs and salad oil) in place of the raw egg.

Cooling, Storage and Reheating

► Potentially hazardous foods containing cooked eggs must be stored in the refrigerator or freezer as soon as possible. If these foods have been allowed to remain in the danger zone for more than two hours, they should be thrown out.

► Refrigerated leftovers should be used within three days and consumed cold out of the refrigerator or heated to piping hot temperatures. Never just "warm up" leftover foods containing cooked eggs and then eat them.

► Make sure these foods are reheated quickly to 170°F (77°C). This will destroy any microbes that have been introduced since the previous cooking stage.

► If you are reheating food for a single or family-size serving, reheat only enough for that occasion and leave the remainder refrigerated. This ensures that bacteria in the food will not have added time to grow during reheating and cooling.

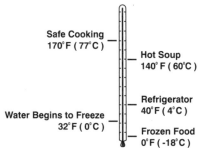

Safe Cooking
170°F (77°C)

Hot Soup
140° F (60°C)

Refrigerator
40°F (4°C)

Water Begins to Freeze
32°F (0°C)

Frozen Food
0°F (-18°C)

Egg Handling — Myths and Facts

Myth: Newly laid eggs are sterile on the inside.

Fact: No raw foods are sterile. *Salmonella enteritidis* bacteria attach themselves to the ova of the chicken and are present in the egg even before it is laid. Reports of illness from this strain of bacteria are on the increase around the world.

Myth: Eggs don't require refrigeration.

Fact: Eggs are a perishable and potentially hazardous food. They must be held at refrigeration temperatures to slow the growth of any dangerous bacteria that may be on or in them.

The number of *Salmonella* bacteria that form in the egg can increase rapidly outside of refrigeration temperatures. That's why all eggs should be refrigerated continuously—from the farm—to the kitchen—to the time they are cooked.

Myth: Eggs produced from chickens that are raised "organically" don't carry any harmful bacteria.

Fact: Harmful bacteria can be present on or in any poultry or meat product. Bacteria are a natural part of our environment, and *Salmonella* bacteria can survive anywhere the conditions are right for them.

Cereals and Grains

These foods are considered safe in their dry form because moisture—one of the important conditions a microbe needs to survive—is not present. Uncooked cereal and grains may contain spores produced by bacteria when the food was dried.

Since our kitchens lack labs to test for microscopic spores, we must assume these foods are contaminated. Handle them appropriately.

The spores may not be destroyed by boiling the food. Many spores, in fact, need to reach temperatures well above boiling before they are destroyed. When we cook these foods at normal cooking temperatures, we provide the conditions the spores need to vegetate and grow—i.e. we add moisture and soften the hard foods, making it possible for the bacteria to use them.

Bacillus Cereus bacterial spores are often carried on raw rice and have been the source of foodborne illness. Because the bacteria had access to the nutrients, moisture, neutral pH and were allowed to remain in the danger zone temperature range, the spores survived the cooking step and were able to vegetate and grow in the cooked rice.

If we do not keep these cooked foods refrigerated after cooking (especially rice), or above 140°F (60°C), the spores will have everything they need to germinate, grow and multiply. Within two hours at room temperature, there can be enough bacteria or toxins to cause illness.

Purchasing

► Always buy cereal and grains from a reputable source.

► Check the packaging for tears or other signs of damage or molds. If moisture is allowed to contact these products, microbes could have the conditions they need to grow. Damaged packages could also indicate contamination from insects or rodents.

► Reject any of these foods if they have signs of mold.

► It is always best to buy the freshest product possible.

Storing

► Cereal and grains should always be stored in a dry cool area away from sunlight. Light can cause photodegradation by oxidizing the fats in these foods. Oxidized fats are thought to be carcinogenic. Most people will not eat foods that have a rancid flavor, but some people do not seem to be affected by the change in flavor.

► Placing these foods in a sealed container will help prevent exposure to moisture or insects and pests.

► Whole, dried, rice and grains can be stored for several years as long as no moisture comes in contact with them.

► Processed, raw grains such as rolled oats can be stored for up to one year.

► Processed, cooked and dried grains such as cereal flakes can be stored for eight months.

► Wheat bran and wheat germ can be stored in a cool area for six months or in the freezer for one year.

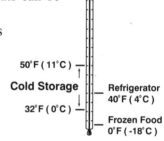

50°F (11°C)

Cold Storage

32°F (0°C)

Refrigerator
40°F (4°C)

Frozen Food
0°F (-18°C)

Preparation

► Rice and other whole grains require a thorough rinse in cold water to remove dirt and debris that may contain pathogens.

► If your recipe calls for a soaking period longer than one hour (i.e. for tabouli or a whole grain cereal) it should be done in the refrigerator.

Cooking

► Rice and whole or crushed grains are brought to a boil and then simmered for a time so they are soft enough for us to eat. If the food will be simmered for more than one hour, the temperature should stay above 140°F (60°C).

► To be sure all parts of the food stay hot enough to prevent bacteria from growing, stir simmering foods often.

Cooling, Storage and Reheating

▶ Once cereal and grains have been cooked, they are soft, moist and nutritious so any bacteria that are introduced will now have the conditions they need to grow. Any spores that were in the dried product can vegetate and grow cells.

▶ Cereal and grains become hazardous foods after cooking, and they must be stored in the refrigerator or freezer as soon as possible. If these foods have been allowed to remain in the danger zone for more than two hours, they should be thrown out.

▶ Refrigerated leftovers should be used within three days and consumed cold out of the refrigerator or heated to a piping hot 170°F (77°C). Never just "warm up" leftover foods containing cereal or grains and then eat them.

▶ If you are reheating food for a single or family-size serving, reheat only enough for that occasion and leave the remainder refrigerated. This ensures the bacteria that may be in the food will not have added time to grow during reheating and cooling.

Cereal and Grains — Myths and Facts

Myth: Foods like cooked rice or oats do not need to be refrigerated quickly. Because they are not meat or dairy products, nothing dangerous can grow in them.

Fact: Once cereals and grains are cooked, they become perishable foods. They must be held at refrigeration temperatures to slow the growth of any dangerous bacteria that may be in them. Remember that microbes are everywhere and they will grow in any food that provides the conditions they need.

Myth: Spicy foods like falafels or grains that are marinated, as in tabouli, are free of microbes because the neutral pH is altered by lemon juice or spices.

Fact: The acid in lemon juice will be diluted by the fluid used to reconstitute the grains. Spices like cayenne taste hot to us, but they have no influence on microbes. The only "heat" that will destroy the microbes is thermal heat of 170°F (77°C).

Legumes (beans, peas and lentils)

As with grains, when these foods are in their dry form, they are considered safe because one of the important conditions that microbes need to survive is not present—moisture. Uncooked legumes may contain spores produced by bacteria when the food was dried.

Because we have no way of detecting these microscopic spores in our kitchens, we must assume these foods are contaminated and handle them accordingly.

The spores may not be destroyed by boiling the food. Many spores need to reach temperatures well above boiling before they are destroyed. When we cook these foods at normal cooking temperatures, we provide the conditions the spores need to emerge and grow —i.e. we add moisture and soften the hard foods, making it possible for the bacteria to use them.

If we do not keep these cooked foods refrigerated or above 140°F (60°C) the spores will have everything they need to emerge, grow and multiply. Within two hours at room temperature, there can be enough bacteria or toxins to cause illness.

Purchasing

▶ Always buy legumes from a reputable source.

▶ Check the packaging for tears or other signs of damage. If moisture is allowed to contact these products, microbes could have the conditions they need to grow. Damaged packages could also indicate contamination from insects or rodents.

▶ Reject any of these foods if they have signs of mold.

▶ It is always best to buy the freshest product possible.

Storing

▶ Legumes should always be stored in a dry cool area away from sunlight. Light can cause photodegradation by oxidizing the fats in the food. Oxidized fats are thought to be carcinogenic. Most people will not eat foods that have a rancid flavor, but some people do not seem to be affected by the change in flavor.

► Placing these foods in a sealed container will help prevent exposure to moisture or insects and pests.

► Dried whole legumes can be stored for several years as long as no moisture comes in contact with them.

Preparation

► Dried legumes require a thorough rinse in cold water to remove dirt and debris that may contain pathogens.

► Many legumes require a soaking period, usually overnight, to tenderize them and reduce the cooking time. These foods are safe to soak in cold water at room temperature for up to 12 hours because they are in their raw state.

Cooking

► Most dried beans, peas and lentils take a long time to cook. They need to be brought to a boil and then simmered for several hours. If the food will be simmered for more than one hour, the temperature should stay above 140°F (60°C).

► Most recipes that have legumes as a base, have many other ingredients added during the cooking process. Each time you add new ingredients, it is very important to bring these foods back to at least 170°F (77°C).

► But, by bringing these foods back to a boil 212°F (100°C) each time you add new ingredients, (including vegetables, meats or fresh spices) you can be sure you've destroyed any pathogens that are on them.

► To be sure all parts of the food stay hot enough to prevent bacteria from growing, stir simmering foods often.

► By stirring frequently, you will distribute oxygen as well as heat through the food. The oxygen and heat will prevent any *Botulinum* spores that may have been introduced on vegetables like onions or garlic from growing.

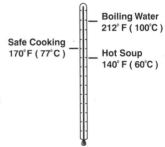

Boiling Water
212° F (100°C)

Safe Cooking
170°F (77°C)

Hot Soup
140° F (60°C)

Cooling, Storage and Reheating

► Once legumes foods have been cooked they are soft, moist and nutritious so any bacteria that are introduced will now have the conditions they need to grow. Any spores that were in the dried product can emerge and grow.

► Legumes become potentially hazardous after cooking and they must be stored in the refrigerator or freezer as soon as possible. This includes tofu and soy milk beverages.

► Some foods made with legumes can be very dense for example, refried beans or chili. In order to cool them quickly enough to keep microbes from growing in them, it may be necessary to spread them in a shallow pan.

► It can take several hours to cool batches of refried beans if they are more than three inches deep. That much time can allow the dangerous microbes to reach levels high enough to make us sick if we eat the food.

► If these foods have been allowed to remain in the danger zone temperature range for more than two hours, they should be thrown out.

► Refrigerated leftovers should be used within three days and consumed cold out of the refrigerator or heated to piping hot temperatures. If your recipe used many ingredients, you are best to use the leftovers within two days. Never just "warm up" leftover legumes and then eat them. It could be hazardous to your health.

► Make sure these foods are reheated quickly to 170°F (77°C). This will destroy any microbes that have been introduced since the previous cooking stage.

► If you are reheating food for a single or family-size serving, reheat only enough for that occasion and leave the remainder refrigerated. This ensures that the bacteria that may be in the food will not have added time to grow during reheating and cooling.

Safe Cooking 170°F (77°C)

Hot Soup 140°F (60°C)

Refrigerator 40°F (4°C)

Water Begins to Freeze 32°F (0°C)

Frozen Food 0°F (-18°C)

Legumes — Myths and Facts

Myth: Foods like beans do not need to be refrigerated quickly because they are not animal or dairy products, so nothing dangerous can grow in them.

Fact: Once beans, peas or lentils are cooked, they become perishable foods. They must be held at refrigeration temperatures to slow the growth of any dangerous bacteria that may be in them. This includes tofu and soy milk beverages which are bean products.

Remember that microbes are everywhere and will grow in any food that provides the conditions they need.

Myth: Foods like chili and hummus are safe because they have strong spices in them.

Fact: Spices like cayenne taste hot to us, but they have no influence on microbes. The only "heat" that will affect the microbes is thermal heat of 170°F (77°C).

Nuts

Dried nuts add flavor and texture to many favorite recipes. Some are ground to produce spreads like peanut butter or they may be dressed with salt, sugar or other spices and roasted for snacks. Some are eaten right out of the shell.

Serious health concerns have been raised because levels of afla-toxin, thought to be very carcinogenic, have been detected in foods made from nuts that were contaminated with mold. Aflatoxins are produced by certain strains of mold as they grow, and some are not destroyed by high temperatures.

The nuts most commonly associated with these toxins are peanuts, brazil nuts and pistachio nuts. Many countries have estab-lished limits for acceptable levels of this toxin in foods.

Manufacturers of foods such as nut butters or nut products have developed detection systems that allow them to remove nuts that have been affected by mold.

It is more difficult for a supplier to detect the presence of mold on nuts that are still in the shell. Many countries sample nuts in the shell to test for the presence of aflatoxin in a shipment before they are sold for public consumption.

It can be difficult to tell by looking, whether or not a nut that you have cracked has been affected by mold. If there is any discol-oration or shrivelling of the nut don't taste it! If you bite into a nut and it tastes bad, spit it out! The level of aflatoxin in a nut that tastes bad could be much higher than the recommended level.

Purchasing
► Always buy nuts from a reputable source.
► Check the packaging for tears or other signs of damage. If moisture is allowed to contact these products, microbes could have the conditions they need to grow. Damaged packages could also indicate contamination from insects or rodents.
► Reject any of these foods if they have signs of mold.
► It is always best to buy the freshest product possible.

Storing

► Nuts should always be stored in a dry, cool area away from sunlight. Light can cause photodegradation by oxidizing the fats in the food. Oxidized fats are thought to be carcinogenic. Most people will not eat foods that have a rancid flavor, but some people do not seem to be affected by the change in flavor.

► Packaged or unpackaged (bulk) nuts have a shelf life of one to two months. Nuts sealed in jars or cans have a shelf life of one year if they are unopened.

► Placing these foods in a sealed container will help prevent exposure to moisture or insects and pests.

► The manufacturers label will indicate whether or not a nut butter requires refrigeration. Some manufacturers also provide a "use by" date on nut butters.

Fruits and Vegetables

Fruits and vegetables are a substantial part of our diet. World-wide distribution of these foods is common. When these foods are shipped, any microbes present on or in them will be transported with them.

Mold spores are also often carried on the surface of fruits and vegetables. Some molds produce mycotoxins which can form in and around the mold growth. Some of these are highly toxic even in small amounts and, in animal tests, have been found to cause cancer. Some mycotoxins can survive for a long time in food. Some are not destroyed by heat so cooking won't help.

When fruits or vegetables are raw and undamaged, they are too hard for the dangerous microbes to use. Most fruits have high acid levels which create an environment that makes it difficult for dangerous microbes to reproduce rapidly.

Low-acid fruits and vegetables become potentially hazardous when cooked because they become soft and moist enough to allow the growth of microbes.

Potatoes with green discoloration is an indication that they have been stored improperly and exposed to sunlight. This exposure causes the production of Solanine, a naturally occurring toxin. Don't eat potatoes with green skin. This toxin will not be destroyed by cooking.

There have been outbreaks of diseases such as Salmonellosis traced to foods like cantaloupe which carried the bacteria on the outer rind. They were transmitted to the edible surface of the melon by the knife used to cut it and grew to dangerous numbers because it was held at cool, but not cold, temperatures in a salad bar.

These foods can become contaminated by workers hands, contaminated irrigation water, septic systems breaking down (during a flood) or through contamination during shipping.

There have also been cases of Salmonellosis food poisoning from alfalfa sprouts when the seeds were contaminated with the bacteria. The best defense against these situations is to buy your food from a reputable supplier—someone who demonstrates an understanding of safe food handling practices.

Cooked Salads

Foods like potato salad, or any other salad that contains cooked vegetables, meats or seafoods, are potentially hazardous because they are often mixed with other ingredients, such as dressing and eggs, that provide the things pathogens need most. Because we think of vegetables as being safe, we may not realize how they have changed through cooking.

Salads made from pre-cooked ingredients must be prepared carefully to protect them from cross-contamination. They must also be kept at refrigeration temperatures of 40°F (4°C) until used. They must not be held at room temperature for more than an accumulated two hours. This includes preparation time.

Botulism and Vegetables

Botulinum bacteria form spores when the conditions they need are not present. These spores act like tiny seeds waiting until conditions change so they can germinate and grow. They are often found on any food that comes in contact with the soil such as root vegetables, herbs and salad greens.

The spores are not destroyed through normal cooking because it takes a very high temperature for a sufficient time to destroy them. That is why a pressure cooker must be used when canning vegetables and other low-acid foods. The pressure canner produces temperatures above boiling 212°F (100°C) to at least 240°F (116°C).

— Pressure Canning
240° F (116°C)

— Boiling Water
212° F (100°C)

Safe Cooking —
170°F (77°C)

Because the high acid levels create an unsuitable environment, the bacteria will not germinate from the spores inside the stomach of a normal healthy person.

These particular bacteria require an absence of oxygen to grow. That is one reason why they can be found in improperly processed, home-canned foods. But, these bacteria can emerge in some surprising places if the conditions they need are present.

The toxin produced by *Botulinum* bacteria is considered to be one of the deadliest toxins known.

Garlic or Herb Oils

Because oil creates an oxygen-free environment, home-made flavored oils can provide the right conditions for *Botulinum* spores to vegetate and grow. Since these oils are often flavored with fresh garlic or herbs, which come in contact with soil, the spores are often present in them.

A recipe for flavored oil brings the bacterial spores into the oxygen-free environment they need to vegetate. If the flavored oil is not kept refrigerated, there is enough moisture and nutrients in the fresh garlic and herbs to support growth and toxin production.

We are now aware of some strains of *Botulinum* that continue to grow and produce toxins at refrigeration temperatures. And so, it is recommended that home-made flavored oils be used at the time they are prepared, or stored under refrigeration and used within two days.

Some commercially-produced oils are treated to destroy the *Botulinum* spores and may not require refrigeration. Others may not be treated, so be sure to read the label.

Foil-baked Vegetables

There have been cases where *Botulinum* bacteria have caused serious illness because the bacteria grew and formed toxins in foil wrapped baked potatoes. The foil was wrapped tightly enough to create an oxygen-free environment. When the potatoes were cooked, the oxygen was driven from the potatoes leaving the food soft and moist enough for bacteria to grow in. The spores germinated and formed the toxin after a few days in unrefrigerated storage.

If the foil had been removed, the oxygen in the air would have prevented *Botulinum* bacteria from growing. Allowing oxygen to get to the food will prevent Botulism toxins from being formed, but the foods still need to be kept hot —above 140°F (60°C), or cold— below 40°F (4°C) to prevent other microbes from growing in them.

Purchasing

► Always buy fruits and vegetables from a reputable source.
► Because these foods are often eaten without cooking them, there is no opportunity to kill any pathogens that may be on them. So, it is important to purchase good quality produce with no visible signs of spoilage.

► Enzymes are proteins found in plants and animals. Enzyme action causes uncooked foods to break down and creates undesirable changes in color, taste and texture. Spoilage organisms may cause illness in some people.

► Reject fruits and vegetables if they have signs of mold. Recall that mold is like a plant. There can be contamination throughout the food because what we see is only the "flower." There is a stem and root system penetrating the food as well.

► Cutting out the spots you can see may not be enough to make it safe. If mycotoxins have had a chance to form, they will be present in the food. You would need to cut one inch around and under a mold or "rot" spot to remove the contamination, and that should only be done if the affected area is smaller than the size of your fingernail.

► In fact, it is recommended to not open a plastic bag once mold is visible on the fruits or vegetables inside. Not only can the spores contaminate your kitchen, our bodies can provide the environment the pathogenic molds need to grow. Spores can germinate on the surface of the eye, causing blindness, or in the lungs, causing respiratory problems.

► It is always best to buy the freshest produce possible.

Storing

► Fruits and vegetables should always be stored in a dry cool place or in the refrigerator.

► Cardboard is not recommended for long term storage because it can harbor moisture and support the growth of molds or bacteria.

► Spoilage organisms can spread rapidly if they come in contact with other foods. It is important to check frequently for fruits or vegetables that are beginning to spoil so you can remove them and protect the rest of your food.

► Potatoes must be stored in a dark area or a heavy cardboard box. Exposure to sunlight causes the production of Solanine toxin.

► When your refrigerator is operating at the correct temperature of 40°F (4°C) or colder, fruits or vegetables that contain
a lot of water (i.e. lettuce), may freeze if the temperature
approaches 32°F (0°C). These foods will be better protected
from temperature fluctuations and excessive moisture loss in
the crisper drawers of the refrigerator.

Preparation
► Always wash or scrub the skin or rind of fruits and vegetables with cool or warm water before using them.
► If you have prepared a dish with raw fruits or vegetables,
such as a salad, it is important to
make certain that it doesn't come
into contact with with other raw
foods, in particular, meat, poultry
and fish products. Uncooked foods can
transfer harmful microbes to raw fruits and
vegetables and there will be no cooking step to
destroy them.

Cooking
► When cooking low-acid fruits or vegetables, it is best to
bring them to a high enough temperature 170°F (77°C) to
kill any pathogens that may be on them. If the recipe calls
for a long simmering period of one hour or more, the
temperature should stay above 140°F (60°C).
► Stir simmering foods often, to be sure all parts of the food
stay hot enough to prevent bacteria from growing.

Cooling, Storage and Reheating
► Once low-acid fruits and vegetables have been cooked they
are soft, moist and nutritious so any bacteria that are introduced will now have everything needed to grow. Any spores
that were on the raw product can germinate and grow.
► These low-acid foods are now potentially hazardous, and
they must be stored in the refrigerator or freezer as soon as
possible.

► If cooked, low-acid fruits and vegetables have been allowed to remain in the danger zone for more than two hours, they should be thrown out.

► Refrigerated leftovers should be used within three days and consumed cold out of the refrigerator or heated to piping hot temperatures. Never just "warm up" leftover, cooked vegetables and then eat them.

► Make sure these foods are reheated quickly to 170°F (77°C). That will destroy any microbes that have been introduced since the previous cooking stage.

► If you are reheating food for a single or family-size serving, reheat only enough for that occasion and leave the remainder refrigerated. This ensures that any bacteria that may be in the food will not have additional opportunities to grow.

Mushrooms

Mushrooms are an edible fungus that is often served or prepared in the same way we use vegetables. Many species of mushrooms can be extremely toxic to humans and animals. Most people are aware that some mushrooms are not safe to eat. Yet, each year there are cases where people have become seriously ill or died because they have mistaken a poisonous mushroom for an edible one.

► Only properly trained people should harvest wild mushrooms because many poisonous mushrooms look very similar to edible varieties.

► Always buy mushrooms from a reputable supplier.

► Keep mushrooms in a paper bag or a wrapper with holes that will allow air to reach the mushrooms. This will prevent the development of slime that occurs when mushrooms are held in an airtight container or plastic bag.

► If mushrooms are held in an airtight plastic bag or container for several days without refrigeration, they can support the growth of *Botulinum* bacteria which produce the deadly Botulism toxin.

▶ Mushrooms contain a lot of moisture. When your refrigerator is operating at the correct temperature of 40°F (4°C) or colder, they may freeze if the temperature approaches 32°F (0°C). They will be better protected from temperature fluctuations and excessive moisture loss in the crisper drawers of the refrigerator.

Fruits and Vegetables — Myths and Facts

Myth: *Low-acid fruits and vegetables are safe because they don't have enough protein to support bacteria growth.*

Fact: After cooking, these foods can provide the right environment for bacteria and other pathogens to grow rapidly. They must be thrown out if they are held at room temperature for more than two hours.

Any damage to the outer surface of raw fruits or vegetables— such as cuts, punctures or damage from enzyme activity or molds— makes them more susceptible to invasion by microorganisms.

Viruses like *Hepatitis A* and bacteria like *Salmonella, Listeria* and *E. coli* can be carried on raw fruits and vegetables including lettuce, cucumber, peppers (capsicums) and others. These foods can be contaminated in the field, in transport or during preparations.

Myth: *All fruits have high acid levels and so the pathogens cannot grow in them.*

Fact: Some foods we think of as fruit, like melons, do not have high acid levels. Through greenhouse technology, some foods that we think of as high-acid foods—like some varieties of tomatoes— are now grown with lower acid levels.

An outbreak of *E. coli* has been linked to unpasteurized apple cider. Raspberries have been implicated as a source of the *Hepatitis A* virus, and associated with an outbreak of the *Cyclospora* parasite. These foods were probably contaminated in the field, during transport or in processing. Fruit juices can also support the growth of yeasts and molds.

Beverages

Many beverages are stored in the refrigerator because they taste better to us and retain vitamins longer when they have been kept cold. However, some beverages (i.e. soy milk and vegetable juice) must be stored at refrigeration temperatures to keep them safe.

With the new technologies available for packaging foods and beverages, come changes in the way these products can safely be stored. We can't rely on "how we've always" handled foods in the past.

Always read the labels on these foods and follow the producers instructions for refrigeration. Some labels will say "keep refrigerated," which means refrigerate at <u>all</u> times—even before you open it. Many others will say "refrigerate after opening." If there are no instructions for refrigeration, the product should be safe at room temperature.

Dairy Beverages

All dairy beverages must be held at temperatures either cold or hot enough to keep pathogens from multiplying rapidly in them. These are high protein, moist, low-acid foods that provide the right conditions for microbe survival and growth.

Commercially-sold dairy products have a "best before" date on the container. If you leave these products out of the refrigerator for any amount of time the product will have a shorter life.

Beverages made from dairy products like milk, buttermilk or chocolate milk should be thrown out if they have been allowed to remain at room temperature for more than a total of two hours.

Because some pathogens, like *Listeria*, can multiply at refrigeration temperatures, it is important to protect these foods from cross-contamination.

Unpasteurized dairy products have been the source of many cases of serious foodborne illness. Using unpasteurized dairy products in any form is very risky.

Raw Egg in Beverages

Beverage recipes that call for raw eggs—i.e. egg nog or "athlete's power drinks"—are no longer recommended. There is the danger that *Salmonella* or other microbes from the surface of the eggshell (they are often contaminated with waste material from the hen during the laying process) could contaminate the drink.

In addition, the strain of *Salmonella enteritidis* has been known to attach itself to the ova of the chicken. So, it could be inside the egg even before the egg is laid. We have no way of knowing which egg may contain this strain of the bacteria, and so we have to assume that all eggs carry this bacteria.

Eating raw eggs is not recommended

It is possible to purchase pasteurized eggs in a variety of forms. These are the only uncooked eggs that can safely be added to beverages that won't be heated.

There are recipes for egg nog that call for cooking the eggs as a custard and then mixing that with cream. This is a safe method for preparing home-made egg nog because the eggs are cooked to a temperature that will kill *Salmonella* bacteria.

Fruit and Vegetable Juices

Beverages made from fruit or vegetable juice must be refrigerated after opening, mixing from a concentrate, or juicing from fresh produce. They are perishable foods because they can support microbial growth or carry pathogens.

Home-made Juice

When processing juice from fresh produce, it is very important to use undamaged fruit or vegetables. Using produce that has any sign of spoilage is risky. The juice could contain pathogenic microbes that were on the raw product.

It's not safe to use up old or damaged produce by juicing it. These foods are susceptible to invasion by pathogens once their outer surface or tissue is damaged. Even cutting away visibly spoiled areas may not remove all of the contamination.

Don't use produce that has signs of mold. Heat doesn't destroy the mycotoxins produced by some molds, and they can be present throughout the food.

Desserts

Many desserts are considered to be safe foods because they contain essentially dry ingredients and enough sugar to prevent pathogenic microbes from growing quickly.

Foods containing dairy products like whipped cream, sour cream, yogurt, or custard are potentially hazardous and need to be refrigerated. These foods should be thrown out if they are left at room temperature for more than two hours.

Cakes and Cookies

Cakes and cookies are considered to be safe foods after they have been cooked. The cooked product is dry and there is enough sugar to inhibit the growth of dangerous pathogens.

If they don't contain potentially hazardous foods such as dairy fillings, cakes and cookies can be safely stored at room temperature for several days. After cooling, they need to be kept covered, to prevent contamination from people, dust, or insects such as flies.

Preparation

► It is important to prevent cross-contamination of work surfaces and tools from ingredients like raw eggs when mixing batter for cakes or cookies.

► Tasting uncooked cake or cookie dough is a risky practice if the dough contains raw eggs. The raw eggs may contain *Salmonella* bacteria and cause foodborne illness. This illness can be serious for children, ill or elderly people who are at high risk from food poisoning.

Fruit Filled Desserts

Desserts made with fruit fillings have a high sugar content, and in most cases, high acid levels. These foods can be safely stored at room temperature for several days. Keep them covered after cooling, to protect against contamination from people, dust or insects.

Selecting Fruits

► Enzymes are proteins found in plants and animals. Enzyme action causes uncooked foods to breakdown and creates undesirable changes in color, taste and texture. Spoilage organisms may cause illness in some people.

► Reject any fruits if they have signs of mold or spoilage. Mold is like a plant. There can be contamination throughout the food because what we see is only the "flower" and there is a stem and root system penetrating the food as well. Many types of mold produce mycotoxins as they grow.

► It may not be enough to cut out the moldy spots you can see. If mycotoxins have had a chance to form, they may be present throughout the food. You would need to cut one inch around and under a mold or "rot" spot to remove the contamination, and that should only be done if the affected area is smaller than the size of your fingernail.

► It is always best to use the freshest produce possible.

Preparation

► Always wash or scrub the skin or rind of fruits with cool or warm water before using them.

Custards in Desserts

Custards made with dairy products are moist, soft, protein rich foods. These foods provide an excellent environment for harmful bacteria to grow whenever they are held in the danger zone temperature range. After cooking, custards must be cooled quickly to slow the growth of harmful bacteria.

Some custard recipes call for raw egg whites to be mixed into the base ingredients, which have been cooked. Since the raw eggs may carry harmful bacteria, they can contaminate the custard. It is safer to use a recipe that includes further cooking of the egg whites or, use pasteurized egg products as an alternative.

Cooling

► To cool the custard rapidly, place it in a shallow dish and stir it as it cools. Metal dishes work best because metal is a good conductor to transfer the heat from the custard.

▶ By using a shallow dish you create more surface area for the heat to escape from. Stirring also helps the heat to escape more quickly.

▶ You can speed up this process by placing the dish on a bed of ice or setting it in cold water. Cooling wands (food grade plastic, gel-filled wands that are stored in the freezer until needed) can be used as well.

▶ Don't use plastic dishes to cool hot food. Some plastics are not designed to hold hot foods.

▶ If the custard is to be used as a filling in a pie, place the cooled custard in a pre-cooked pie crust to speed up the cooling time.

▶ Custards and pies made with custards should be held in the refrigerator. If they have been allowed to remain at room temperature for more than two hours, they should be thrown out.

Dairy Topped or Filled Desserts

Because they contain dairy products, cream or yogurt topped or filled desserts must be held at refrigeration temperatures. This is still the case when they have been added to a dessert, such as carrot cake, that would be considered safe at room temperature. Dairy topped or filled desserts should be thrown out if they have been allowed to remain at room temperature for more than two hours.

Dairy products provide the rich, moist and neutral pH foods that bacteria need to reproduce and survive. If these foods are not kept at refrigeration temperatures of 40°F (4°C) or colder, microbes in the foods will have the right conditions to grow.

For example, each time you leave a container of cream on the kitchen counter, any spoilage microbes that may be in the cream will reproduce more quickly and that would affect the validity of the "best before" date.

Preparation

NOTE: Most dairy products are "ready-to-eat" foods. Care should be taken to prevent cross-contamination from raw foods or work surfaces and utensils. Keep a separate work area or cutting board and tools for ready-to-eat foods.

Cooking and Mixing

► Dairy products are often served without ever cooking them at a high temperature. Because there is no added "kill temperature" in the process, we need to pay close attention to how we store, handle and prepare them.

► When preparing foods that contain dairy products plan your steps in a way that will keep these foods at refrigeration temperatures as long as possible.

► If your recipe includes sour cream which is mixed with other ingredients at the end of the process, leave the sour cream in the refrigerator until the moment it is needed. Then use what you need and put it back in the refrigerator.

Hot Soup
140° F (60°C)

Body Temperature
98.6°F (37°C)

Danger Zone

► For example, if you were to leave sour cream on the counter for 45 minutes, the two-hour safety rule is reduced to 1 and 1/4 hours for the remaining sour cream and your creation as well.

Refrigerator
40° F (4°C)

Eggs in Desserts

Many dessert recipes include eggs. In some traditional recipes the eggs are not cooked.

When an egg is laid it can become contaminated with waste material from the hen. If the hen is carrying *Salmonella* bacteria, it will be passed to the eggshell. Some commercial egg producers wash the eggs to remove the waste material, but bacteria can still be present.

The *Salmonella enteritidis* strain of bacteria can be found in the ova of the hen. This means that some eggs can be formed with the bacteria inside the shell, even before they are laid. Reports of illness caused by this strain of *Salmonella* are becoming more widespread around the world each year.

Eating raw eggs in any form is no longer recommended. Use pasteurized egg products as an alternative.

A wide variety of pasteurized egg products are currently available for restaurant and institutional use. Some products are now available in retail food stores, for home use. If they are not available at your local store, ask the manager to order them.

Eating raw eggs is
not recommended

Preparation

NOTE: Any surface that comes in contact with raw eggs or eggshells can be contaminated with the bacteria that are hitch-hiking on or in them.

▶ To prevent cross-contamination, keep a separate work area and tools for preparing or handling raw eggs. This also makes it easier to clean all possible contact surfaces when you've finished working.

▶ After you have finished preparing the raw eggs, all work surfaces, dishes and tools should be washed and then sanitized using a solution of water and bleach. The recommended ratio is one ounce (30 ml) of bleach per gallon (4 L) of water.

▶ Raw cake batter, cookie batter, or any other food that contains uncooked eggs should not be tasted before cooking because it could contain *Salmonella* bacteria.

▶ Some recipes call for raw whipped egg whites to be added to the finished dish. Eating raw eggs in any form is not recommended. Use pasteurized egg products instead.

▶ Desserts made with pasteurized egg products still need to be held at refrigeration temperatures of 40 °F (4°C) or colder. If they have been allowed to remain at room temperature for more than two hours, they should be thrown out.

Meringue Topping

If your recipe calls for a meringue topping, add the meringue to hot filling and bake it until it is thoroughly cooked to 170°F (77°C). This will take approximately 15 minutes at an oven temperature of 350°F (180°C).

Pies topped with meringue are best stored at refrigeration temperatures of 40°F (4°C).

Dairy Substitutes

Because they are made from vegetable and other edible oils, we may not think of dairy substitutes, including soy products, as perishable.

Dairy substitutes should be handled as potentially hazardous foods because they give pathogenic microbes the environment they need to survive. These foods must be refrigerated or frozen. Always read the label and follow the manufacturer's instructions for storing dairy substitutes.

Icing

Icings that contain dairy products, dairy substitutes or eggs should be held at refrigeration temperatures of 40 °F (4°C) or colder. If allowed to remain at room temperature for more than two hours, they should be thrown out.

Breads

Bread, in its many forms, has been a main part of our diet throughout history. It is a versatile and nourishing food for people around the world.

Bread is considered a safe food because baking makes it too dry to support rapid growth of harmful bacteria. After cooling, bread needs to be kept covered to prevent contamination from people, dust or insects. Dust is a potential source of mold spores.

Molds associated with grains have been the cause of illness and many deaths throughout history. Improved growing, harvesting and storage practices have reduced—but not eliminated—the incidence of pathogenic molds in grains and legumes in many countries.

Storing

► When breads are stored in air-proof containers like plastic bags, the moisture that builds up can allow mold spores to vegetate and grow. It can happen more quickly if the bread is stored in direct sunlight or at very warm temperatures.

Mold & its spores

► Don't eat bread with signs of mold. Remember that mold is like a plant and what we see is only the flower. There is a stem and root system penetrating the food, and there can be contamination throughout the loaf. Many types of mold produce mycotoxins as they grow.

► Cutting out the spots you can see may not be enough to make it safe. If mycotoxins have had a chance to form, they will be present in the food.

► While it is true that some molds are beneficial, others can cause serious illness and death. Since we have no way of determining if the mold on our bread is beneficial or pathogenic, it is safest to throw it out.

► In fact it is recommended to not open the bag once the mold is visible. Not only can the spores contaminate your kitchen, our bodies can provide the environment the pathogenic molds need to grow. Spores can germinate on the surface of the eye causing blindness, or in the lungs causing respiratory problems.

Home-Canned Foods

In the early 1800's, foods were preserved by heating them in corked glass bottles. At that time, it was observed that this process prevented the food from spoiling. What they didn't know was why it worked. The microbes that cause food spoilage were not discovered until about 50 years later when Louis Pasteur discovered them with his microscope.

Canisters made of metal have been used since the 1820's. These have been adapted over the years to accommodate new and better processes and to eliminate hazards such as lead soldered seams. Most modern home-canning is done in glass jars with coated metal closures.

In the canning process, foods are heated to a temperature that is hot enough to destroy any microorganisms that may be in the food. During the heating stage, air is forced out of the container. Because the air expands and is driven out, a vacuum forms and draws the sealing lid downward until an airtight seal is created. So air is removed from the container, spoilage organisms in the food are destroyed and no air or microorganisms can get inside the jar.

Ingredients, time and temperature are all important factors in the process for the safe preservation of the food. The level of acidity and the concentration of sugar or salt in the food will determine the correct processing method and time. Some spoilage organisms are inhibited by these factors while others may still grow.

Home-canning has become a very popular way of preserving foods. Unfortunately, mistakes in the processing of low-acid foods can be fatal. The major concern is *Botulinum* bacteria. *Botulinum* can be found everywhere, but especially in meat or fish and on food that comes in contact with the soil—i.e. root vegetables and herbs.

Botulinum bacteria form spores when the conditions they need for growth are not present. These spores act like tiny seeds, waiting until conditions change so they can emerge and grow.

The spores are not destroyed through normal cooking because it takes a very high heat to destroy them. That is why a pressure canner is used when canning vegetables and other low-acid foods. The pressure canner produces temperatures of at least 240°F (116°C) which is well above the boiling point of 212°F (100°C).

Because the high acid levels create an unsuitable environment, the bacteria will not germinate from the spores inside the stomach of a normal healthy person. That is why we don't become ill from eating the spores themselves. It is only when the spores are given the food, environment, temperature and time they need that they will vegetate and grow and form the deadly toxin.

Botulinum bacteria require an absence of oxygen to grow. That is one reason why it is such a concern in improperly processed home-canned foods. The toxin produced by this bacteria is considered to be one of the deadliest toxins known.

High-acid foods (i.e. fruit, pickles, relish, jam, jelly or chutney) can be safely processed at temperatures that destroy the bacteria but not the spores, because the spores will not vegetate in high-acid foods.

When producing canned foods for your family, it is **critical** that the correct procedures are followed and precautions taken. Old fashioned recipes are not always more wholesome. Many people have died as the result of Botulism toxin in improperly home-canned foods.

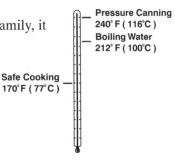

Pressure Canning
240° F (116°C)
Boiling Water
212° F (100°C)

Safe Cooking
170° F (77°C)

Preparation

► Use only high quality foods that show no signs of spoilage. This reduces the risk of contamination from the beginning.

► Follow recipes carefully. The pH or acid levels are an important part of the preservation process. Altering recipes or using substitutions is not a safe practice. Any changes would need to be scientifically tested to ensure safety.

► Use only vinegar of at least five per cent acidity. Do not use vinegar of unknown acidity or home-made vinegar when preserving foods.

► It is important to release air bubbles from the food after packing the jars. Recipes are calculated to process the food long enough to remove oxygen and prevent spoilage microbes from growing in them.

► Once the air bubbles are removed, it is also important to readjust the "head" space to ensure that the air left in the jar will be exhausted during the processing stage.

► **Open kettle or oven canning and recipes calling for wax seals or a boiling water bath for low-acid foods are outdated.** Replace them with modern recipes which follow current procedures.

► **Jars with bailed closures, glass lids or rubber rings are no longer recommended.** Replace them with modern two-piece metal closures which make it possible to easily determine whether or not the seal is intact.

► Use jars designed specifically for home-canning. They are made of heavier glass than jars used for commercial canning, and they can withstand high temperatures and extreme temperature changes.

► If your recipe calls for a processing time as short as five minutes in a boiling water bath, sterilize the jars in boiling water for 10 minutes immediately before filling them.

► Dishwashers, even those with a sanitizing cycle, cannot reach high enough temperatures to sterilize jars that will be used for home-canning.

► If your recipe calls for more than 10 minutes processing time, you can use jars that have only been washed and rinsed. They will be sterilized during processing.

► Two-piece, metal closures with a metal screw band and a metal snap lid provide easy and safe closures for home-canning. As the jar cools, the metal lid is pulled downwards, creating an airtight seal that you can see.

► Because the sealing compound can become too brittle or marked to create a good seal again, metal snap lids can only be used once. The screw bands can be reused until they become damaged by use or rust.

► Follow the manufacturer's directions for handling the metal lids. Some recommend boiling them for five minutes immediately before using them. This sanitizes them and softens the sealing compound so that an airtight seal can form.

▶ Make sure your boiling water canner is large enough to allow one to two inches of water above the top of the jars and one to two inches of room above the water level for a rolling boil.

Processing

▶ The recipe times given for the safe processing of different foods take into account pH levels, density of the food, jar size and the time required to exhaust the oxygen found in each specific combination. Do not substitute any part of a recipe or use a different jar size because this can affect the safe processing time.

▶ Jars of high-acid foods such as jams, jellies, preserves, fruits, pickles, relish and tomatoes with added acid can safely be processed in a boiling water bath.

▶ Jars of **low-acid** foods such as vegetables, meat and game, poultry, fish and seafood, soups and tomato recipes without added acid cannot be safely processed in a boiling water bath. They **must be processed in a pressure canner** to reach the 240°F (116°C) temperature needed to destroy *Botulinum* spores.

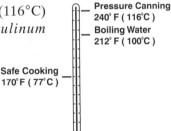

Pressure Canning
240° F (116°C)
Boiling Water
212° F (100°C)

Safe Cooking
170°F (77°C)

▶ Because water boils at lower temperatures at higher altitudes, processing time or pounds of pressure must be increased.

High-acid foods—increase processing time

1,001 - 3,000 feet (306 - 915 meters) 5 minutes
3,001 - 6,000 feet (916 - 1,830 meters) 10 minutes
6,001 - 8,000 feet (1,831 - 2,440 meters) 15 minutes
8,001 - 10,000 feet (2,441 - 3,050 meters) 20 minutes

Low-acid foods—adjust lbs or kPa pressure
(time does not change)

feet	meters	weighted gauge	dial gauge
0 - 1,000	0 - 305	10 lb (68 kPa)	11lb(75kPa)
1,000 - 2,000	305 - 610	15 lb (102 kPa)	11lb(75kPa)
2,000 - 4,000	610 - 1,200	15 lb (102 kPa)	12lb(82kPa)
4,000 - 6,000	1,200 - 1,830	15 lb (102 kPa)	13lb(89kPa)
6,000 - 8,000	1,830 - 2,440	15 lb (102 kPa)	14lb(95kPa)

Cooling and Storage

► Cool the jars for 24 hours. You can often hear the lids of the two part metal closures snap as they are drawn down by the vacuum created as the food cools. Don't disturb the jars by jostling or tilting them until they are cold and sealed.

► Check the seal after the jars have been allowed to cool. The lids on the jars sealed with the two part metal closures should curve downward to show a good seal. If you press your finger on the lid it should not give or spring back.

► If the seals are not good, the ONLY opportunity you have to salvage the food is if you reprocess it immediately (within 24 hours of the original processing). Make sure to sterilize all equipment and use new metal lids. If you choose to refrigerate the food instead, use it within one week.

► Label, date and store your canned foods in a cool, dark, and dry place. The recommended temperatures for cool storage are 32° - 50°F (0° - 11°C).

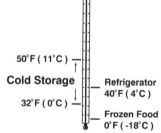

► For best eating quality, it is recommended that home-canned foods be used within one year.

Signs of Spoilage
- NEVER taste canned food that has any signs of spoilage.
- A jar that was sealed and later loses that seal indicates food spoilage. Spoiled food should be destroyed.
- Cloudy liquid, sediment in liquid, gassiness or small bubbles in the food, seepage around the seal, mold around the seal, shrivelled or spongy-looking food, bulging caps or food that is an unnatural color are signs that indicate spoilage.
- Check for spurting liquid or pressure from inside as the jar is opened; fermentation or gas bubbles; soft, mushy or slimy food; an unnatural odor, and any signs of mold.
- But, the Botulism toxin can be present without showing any signs of spoilage, so you may want to take an extra precaution with foods that will be heated before they are served.

Spoiled food must never be eaten!

Botulism Toxin Precaution

If your equipment is in perfect condition and you follow every canning step precisely, including using a pressure canner, you can expect that your low-acid foods will be safe to eat. You can be **certain** that your **perfect** looking jar of low-acid food is safe if you take the extra-ordinary precaution of boiling the food before tasting it. Use a full rolling boil for 10 minutes for home-canned vegetables. Boil corn, meat, poultry and seafood for 20 minutes.

You can only take this precaution with food that **looks perfect** and shows **no signs of spoilage**. Botulism toxin will be destroyed by heating the food to 212°F (100°C) for 10 to 20 minutes, depending on the food.

NOTE: Signs of spoilage can indicate toxins from other sources and those toxins may not be destroyed by heat.

When in doubt—throw it out!

Dry Foods

Many foods are dried for safe storage at room or cool-room temperatures. They contain little moisture so pathogenic microbes can't grow on or in them. However, bacteria or mold spores may be present and they can germinate if these foods get damp or if moisture is added when we prepare them.

The oil in the foods can become rancid because the enzymes that oxidize the fats are still active, even after the food is dried. This rancidity is thought to be carcinogenic and mutagenic. Most people find rancid food unacceptable to taste and some people become ill after eating it.

Purchasing

▶ Always buy from a reputable source.

▶ Dried foods containing natural or added oil, such as crackers or coffee whitener have a six-month shelf life. Other dried foods like ready-to-eat cereals, rolled oats and cornmeal have a six to eight month shelf life.

▶ Some manufacturers include a "best before" date on the packages of dry foods. It is always best to buy the freshest product possible.

▶ Check the package for damage. There could be contamination of the food if there are holes or tears in the packaging.

Storage

▶ Dried foods need to be kept dry. It is best to store them in an area where they will not be exposed to moisture from dish washing, cooking, or steaming kettles.

▶ After opening, dry foods will store best in an airtight container away from light and heat. Light causes photodegradation of the food and so can speed up the action of the enzymes and cause them to become rancid sooner.

▶ Rotate the dried foods in your cupboard by placing the newer ones at the back and pulling the older ones forward. First in, first out.

► If you are stocking a cottage, camper or earthquake food supply, dried goods containing natural or added oils such as crackers, cornmeal, coffee whitener or cereal may become rancid and should be replaced with new stock every six to eight months.

Preparation

► Dried foods are considered safe when they are dry. As soon as we add moisture they become potentially hazardous. For example, when we add water to dry milk powder, it becomes potentially hazardous and must be kept at refrigeration temperatures of 40°F (4°C) or colder.

► When we cook, pasta, rice or beans, they must be kept either hot or cold enough to slow the growth of any pathogens that may be in them through cross-contamination in the kitchen or as a result of spores carried on them from the growing fields.

Commercially Canned Foods

Canning has become a standard way to preserve perishable foods. It is dependable and efficient when done properly. To ensure the quality and safety of their products, manufacturers of commercially canned foods follow strict guidelines and procedures.

The food is processed at high heat to destroy any microbes and their spores and are hermetically sealed against the entry of any pathogens. The sealed environment is the secret to preserving the canned food. If that seal is compromised, the food may become contaminated. Oxygen in driven from the food and the container during the process. This creates an oxygen free environment.

Botulinum bacteria will form spores when the conditions the bacteria need are not present. These spores act like tiny seeds, waiting until conditions change so they can germinate and grow. Spores are often found on food that comes in contact with the soil, such as root vegetables and herbs and also on meat, fish and seafood.

Boiling water is not hot enough to destroy these bacterial spores. That's why a pressure canner or retort system is used. The pressure canner produces temperatures well above boiling—at least 240°F (116°C). The food must be held at these temperatures for a specific amount of time to destroy the spores.

Botulinum spores must be destroyed before they have an opportunity to germinate, grow in the food and produce the toxin that causes deadly Botulism food poisoning. *Botulinum* bacteria require an absence of oxygen to grow. This is the primary reason the toxin could be produced in improperly-processed canned foods.

The bacteria will not germinate from spores inside the stomach of a normal healthy person because the high acid levels in the stomach create an unsuitable environment. So we only have a problem if we eat foods that already contain the toxin.

Purchasing

► Check the can for any abnormal appearance, swells or stains on the label.

► Check that the ends of the can are slightly depressed. If a can bulges, it's a reliable sign that the food it contains is unsafe to eat. If the ends of the can or the lid of the jar are pushing out instead of drawing in, it may be evidence of the deadly toxin produced by growing *Botulinum* bacteria.

▶ If you find a bulging can or a lid that isn't sealed in a store, alert the manager. Commercially canned foods rarely have problems like this, but when it does happen it is taken so seriously that all of the food canned under that particular batch—identified by a code or lot number— will be recalled by the manufacturer or health agency.

▶ Check for dents or deep scratches from knives used to open the shipping carton. These can cause a fracture or opening in the body or seal of the can, which could produce tiny holes. Not only may contamination then get inside the container, any gasses forming inside could escape through these openings. Therefore, the damaged can would not bulge, to indicate the fermentation of gas inside. If this happens, we no longer have evidence that the canned food may be contaminated. It cannot bulge if the seal has been broken.

▶ Any dents along the side seam of the can or the end seams may damage the seal and make the food unsafe. Deep or sharp dents in the body of the can can cause fractures and expose the food to contamination.

▶ Any dent along the score-line of an easy-open, roll top lid could damage the seal and make the food unsafe. The dent could cause a microscopic hole in this vulnerable closure.

▶ Check the can for signs of rust. Rust can cause pitting which can penetrate the metal and destroy the seal.

▶ Check the can for signs of leaking. This may look as though the can or label has had something splashed on it. Several cans on the store shelf may be in this condition because they have come in the same shipment. If the can is leaking, the can itself is breaking down and the seal is not intact. The food inside the can may be contaminated. This situation usually occurs in old stock, or poor quality cans.

Storage

▶ Always follow the manufacturers instructions for storage. **Some canned foods must be kept refrigerated at <u>all</u> times.**

▶ Certain foods, such as anchovies, cheeses or meats are canned using a process that preserves the character of the product where high heat would cause unacceptable changes. Foods processed in this way cannot be safely stored at room temperature.

▶ **Read the labels on all cans before storing them.** Unfortunately, you cannot always depend on the staff at the local grocery store to read the labels and store the foods accordingly. They may have stored the product at room temperature—an unsafe practice for any potentially hazardous food.

▶ Unless otherwise dated, it is recommended that you use canned foods within one year after purchasing them. Many canned goods are labeled with a "best before" date.

▶ Rotate the canned foods in your cupboard by placing the newer ones at the back and pulling the older ones forward to be used first.

▶ If you are stocking a cottage, camper or earthquake food supplies, canned foods should be replaced with new stock each year.

▶ Unless otherwise dated, unopened canned foods that require refrigeration have a four to five month shelf life after purchasing. After opening, they should be used within three days.

Preparation

▶ Before opening, wipe the top of the can with a clean cloth. Remove the label, look for stains on the label and check the body of the can for damage.

▶ When you open the can, you should hear the air being drawn in. If there is any squirting or pressure escaping do not use it. If this happens you should notify the health department so they can investigate the situation. Follow their directions about discarding the contents of the can.

Frozen Foods

Purchasing

▶ Frozen food storage, including open frozen food display areas, must keep these foods frozen solid.

▶ Check for signs that a frozen food has been thawed and refrozen by feeling the food inside the package. For example, if it is slumped into a mass at one end of the package of prepared cabbage rolls or pizza it may have thawed during transport or storage. Producers of these foods go to a great lengths to ensure that the food you buy from them looks perfect when you open it, so the slumping probably occurred somewhere else along the delivery chain.

▶ Moisture on the package or frost inside the package of frozen food are also signs that it may have been thawed and refrozen.

▶ Check the package for damage. There could be contamination of the food if there are holes or tears in the packaging.

▶ Always buy the freshest product possible. Different foods have different storage times based on their size and contents. Foods such as meat, pastry or casseroles containing natural or added oils or fats, have a shorter frozen storage life. The oil in these foods can become rancid because enzymes are still active even when the food is frozen.

▶ Frozen foods should be purchased at the end of a shopping trip and stored immediately in the refrigerator if you intend to use them within the next day or two, or in the freezer if you want to store them longer.

▶ Frozen foods that have been thawed must be cooked thoroughly before they can be safely refrozen.

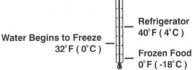

Water Begins to Freeze
32°F (0°C)

Refrigerator
40°F (4°C)

Frozen Food
0°F (-18°C)

Storage

► For long term storage, frozen foods should be stored at 0°F (-18°C) or colder in a chest freezer. The freezer compartment on the top of your fridge may not be as cold as a chest freezer so your stored foods may not maintain the same quality as those stored in a chest freezer.

► Label and date foods that you have prepared or repackaged for freezing.

► Don't create warm spots by placing unfrozen foods close together. Leave space for air to circulate and evenly distribute cold air around foods as they are freezing.

Thawing

► Whenever possible, cook frozen food directly from the frozen state.

► Thaw potentially hazardous frozen foods like meat in the refrigerator. This may take two days and even longer for large items such as roasts. Safe alternatives include microwave ovens, which can do an excellent job if they have a good defrost setting, or running cool water over the item tightly wrapped in plastic followed by immediate cooking.

► NEVER thaw frozen foods at room temperature. Foods that have been frozen are easy for microbes to use. They can multiply rapidly on the warmer surfaces of thawing foods that are still frozen inside.

► Thaw minced or ground raw meats in their own wrapper to prevent cross-contamination. These foods are particularly dangerous because they can contain large numbers of microbes before freezing.

Modified Atmosphere Packaging (MAP)

Newer forms of food packaging are being developed every day—or so it seems. With new technology comes new food handling requirements. As consumers, we can no longer rely on "how we've always done things." We can't even rely on "what makes sense based on what we already know." We MUST read labels for instructions from the manufacturer for proper storage and handling of these "new" packaged foods.

Many people have made the mistake of thinking that when a food is vacuum packaged or in a foil pouch it is safe at room temperature or for a long time. We assume that vacuum packaging or foil pouches have the same effect on food as canning does.

What we may not realize is that the way a food is processed plays an equally important role in how it must be stored. By storing a food in a modified air environment, we may create favorable conditions for pathogenic bacteria that would have been suppressed by other packaging methods—so we need to use other measures to control them.

The **"Keep Refrigerated"** reminder is an increasingly important instruction which is too often ignored. New food production technologies rely on refrigeration as a **critical** safety measure. Many "calorie reduced, sodium reduced, or fat reduced" products must be kept at refrigeration temperatures of 40°F (4°C) or colder and some must be frozen.

Food is packaged in pouches which are flushed with atmospheres different from air before the package is sealed, or the air is removed completely, as in vacuum packaging. Normal spoilage bacteria and fungi are less active in these environments which can extend the usual shelf-life of the food.

Gases which may be used for MAP packaging include:

▶ Carbon dioxide which can change the pH of the food and thereby reduce the growth of some microbes, but will stimulate others—it can also affect the natural color of foods.

▶ Nitrogen, which does not react with food but is part of the gas flushing formula.

▶ Oxygen which is used only in conjunction with nitrogen and/or carbon dioxide.

Bacteria that require air, and mold or fungi, are suppressed by MAP packaging. On the other hand, bacteria that require an absence of air, like *Botulinum,* can be enhanced by MAP packaging unless they are held at refrigeration temperatures of 40°F (4°C) or colder.

When any ready-to-eat fish or seafood is sold in a package that has had the air removed, it must be in a frozen solid state. *Botulinum type E* bacteria are common in the marine environment. These bacteria share the toxin producing abilities of the other strains of *Botulinum* but they may also reproduce at refrigeration temperatures. Another concern for seafood products is *Listeria* bacteria which can also reproduce at cold temperatures.

When fish are frozen, these bacteria cannot reproduce. If you are thawing frozen fish (always done in the refrigerator or microwave) that has been stored in a vacuum package, open the package to allow oxygen to reach the fish and prevent these bacteria from reproducing.

The manufacturer can also make MAP foods "shelf stable" by specialized retorting or other processes, so again we must read labels carefully.

Storage

► Read and follow the manufacturer's instructions for storing MAP foods. Do not guess at correct storage conditions.

► Perishable non-acidic foods that are vacuum packaged in the home must be stored in the same way you would store the raw foods—i.e. at refrigeration temperatures of 40°F (4°C) or colder. Fish or seafood products are an exception because they <u>must</u> be frozen after they have been vacuum packaged.

► Vacuum packaged dry foods, such as coffee, do not require refrigeration because they do not contain enough moisture to support microbial growth.

Water Begins to Freeze
32°F (0°C)

Refrigerator
40°F (4°C)

Frozen Food
0°F (-18°C)

Foods For Special Circumstances

Barbecues

Many people enjoy the flavor of foods cooked on the barbecue, over coals or open flames. We often cook this way at social gatherings. Many cases of food poisoning have been linked to barbecued dinners. This is sometimes referred to as "barbecue syndrome."

Improper handling of food is the main cause of barbecue syndrome. If we're busy enjoying the company of our guests, we may not be paying close attention to the time that foods remain in the danger zone temperature range. Bacteria may be allowed to grow to dangerous numbers in our food.

Cross-contamination of ready-to-eat foods with raw foods has also contributed to these episodes of food poisoning. The same plate may be used to carry the raw food and later used to carry the cooked food. The cooked food can be contaminated by basting it with the marinade from the raw food.

Preparation

NOTE: Any surface that comes in contact with the raw meat or its drippings can be contaminated with hitch-hiking or bathing microbes.

► To prevent cross-contamination, keep a separate work area or cutting board and tools for handling raw meat or poultry. This also makes it easier to be sure you've cleaned all possible contact surfaces when you've finished working.
► Keep foods that are to be barbecued at refrigeration temperatures until it is time to grill them.
► If your recipe requires a marinating period, it should always be done in the refrigerator. Some cooks prefer to leave the item marinating for a longer time when it is refrigerated.

▶ When preparing a cut of **beef or lamb** that will only be cooked to a **medium-rare** stage, or **pork** that will be cooked to a **medium** stage, don't pierce it with a knife or fork. We can expect pathogens to be on the surface of a piece of meat and we do not want to transport them inside the food where the heat may not be high enough to destroy them.

▶ Some recipes advise puncturing meat with a fork to allow a marinade to penetrate the food. This is not recommended if the meat will not be cooked to a well done stage.

▶ If a cut of meat is going to be cooked to a well done stage of 170°F (77°C), any pathogens will be destroyed by the heat. We don't need to be concerned if these foods are pierced during preparation.

▶ After you have finished preparing raw meat, all work surfaces and tools should be washed and then sanitized using a solution of water and bleach. The recommended ratio is one ounce (30 ml) of bleach per gallon (4 L) of water.

▶ Don't forget to wash your hands between handling raw foods and handling ready-to-eat foods.

Cooking

▶ Beef or lamb steaks, chops or ribs should be cooked to a minimum internal temperature of 145°F (63°C). You may chose to cook pork from an inspected source to a medium stage of 160°F (71°C). By the time the center reaches that temperature, the outer surface will be hot enough to kill any bacteria that may be present.

▶ If you intend to cook **beef or lamb** steaks, chops or ribs to a **medium rare** stage, or **pork** to a **medium** stage, use tongs or a flipper to turn them. A fork may transport harmful bacteria from the surface to the center where the temperature will not be high enough to destroy them.

▶ Ground meat patties should be cooked to a well done temperature of 170°F (77°C). Because patties are too thin to actually measure them with a thermometer, they should be cooked until the juices run clear. Cut them open to see inside: **if they are still pink, they need to be cooked longer—especially if the patties are to be eaten by small children.**

► It is recommended that any cuts of wild game, or rolled or stuffed beef, lamb or pork, be cooked to an minimum internal temperature of 170°F (77°C). These temperatures will destroy any pathogens that may be in the meat.

Safe Cooking 170°F (77°C)

Boiling Water 212° F (100°C)

Hot Soup 140° F (60°C)

► If you want to baste the food with a leftover marinade, it should first be brought to a boil to destroy any pathogens that may have been in the raw food. It is important to protect your cooked food from contamination from raw food.

► If you use a brush to baste raw meat, poultry or seafood at the beginning of your cooking period, the brush becomes contaminated with pathogens from the surface of the raw food. If you baste the cooked, ready-to-eat food with that same brush, you will re-contaminate the cooked food with the pathogens you picked up on the brush.

► Partial cooking followed by a finishing stage at a later time is a risky practice. Food at that temperature may feel hot when we touch it. But, in fact, partial cooking means we have only warmed the inside of the food to the microbes' favorite temperature and softened it making it easier for them to use. The cooling and reheating process allows the time the for bacteria to grow to dangerous levels.

► If ribs are parboiled, before they are cooked on the barbecue, they should be grilled immediately after boiling. Don't let them "rest" in the danger zone temperature range.

► If a piece of poultry is too large for the heat of the grill to penetrate through the meat without burning the outer surface, it should be precooked and immediately finished on the barbecue.

► The cooked foods can be contaminated by drippings on the platter that was used to carry the raw foods to the barbecue. Use a clean platter for carrying cooked foods from the grill.

Steaks, Chops and Burgers Touch Test

Use the back of a teaspoon or fork
(Times could change if there is a wind or cool temperatures
and you are cooking out of doors)

MEDIUM - RARE meat gives easily when touched. A hint of juice appears on the surface.
(Minimum four minutes per side for one-inch thick)

MEDIUM meat feels firmer but slightly springy, and juices begin to appear on the surface.
(Minimum Five minutes per side for one-inch thick)

WELL DONE meat is covered with juices. It is firm to the touch and does not yield to pressure.
(Minimum six minutes per side for one-inch thick)

Outdoor Parties

All of the considerations taken into account for barbecues as well as luncheons and dinner parties also apply to outdoor parties. We must pay extra attention to the time that potentially hazardous foods spend outside of safe hot holding or refrigeration temperatures. In addition, weather is generally warmer when we hold parties outdoors.

The "Danger Zone" is a range of temperatures where bacteria will multiply rapidly in food. If we recall that they multiply most rapidly at 98°F (37°C), we can see that a warm summer afternoon will create a very favorable environment for bacteria to quickly grow to dangerous numbers in our food. In an outdoor setting, there is often more opportunity for contamination of the food by dust or insects. Foods should be wrapped or covered to protect them from becoming contaminated.

Body Temperature
98.6°F (37°C)

Hot Soup
140° F (60°C)

Danger Zone

Refrigerator
40°F (4°C)

Luncheons and Dinner Parties

Anyone who has prepared food for a gathering of people knows that it takes good planning and organization to make everything run smoothly. Several dishes may be prepared in advance, some just before guests arrive and still others just before serving.

Refrigerator, range top or oven space may be difficult to find as serving time approaches. It may "seem" necessary to stretch the time and temperature rules by storing some foods in cool rooms or on the back of the range. Food safety rules are not made to be broken. Hot foods must be kept hot and cold foods must be kept cold until they are served.

If the pathogens are allowed to grow rapidly in foods served to a large number of people, a large number of people can become the victims of foodborne illness. Keep in mind that at a large gathering there may also be people at high risk from food poisoning such as the young, the elderly, or someone who has recently been ill.

Preparation

► Plan your menu taking into account available cold and hot holding equipment or space for potentially hazardous foods.

► Providing ice for glasses of soft drinks, or a picnic cooler with ice for canned and bottled drinks, can free up refrigerator space for potentially hazardous foods such as potato and pasta salads or for holding steaks until it is time to grill them.

► If you have a chest cooler, you could use it with ice or freezer packs for overflow from your refrigerator.

► Hot foods need to be held above 140°F (60°C) until they are served. It is best to prepare them as close to serving time as possible so they won't dry out at safe holding temperatures.

► If foods that are to be served hot are prepared in advance and refrigerated before reheating, they should be brought to a "piping hot" 170°F (77°C), and then held above 140°F (60°C) until they are served.

Serving

We often provide appetizers for guests to enjoy while they mingle before the main meal.

Sometimes, there may be a delay in serving the main meal due to late arrivals or a main course—for example, a large turkey— taking longer to cook than was estimated.

When you have several tasks to complete it is important to plan with the safety of the food in mind.

▶ If you are serving appetizers with dips that require refrigeration after opening, such as salsa or home-made dips containing dairy products, set out small amounts and keep the remainder refrigerated until needed.

▶ Dips should be used within two hours after they are removed from the refrigerator. Any remaining dip that is not used in that time should be thrown out.

▶ If you have trays of cold cuts, cheeses or any potentially hazardous food that is not eaten before the main meal is served, they should be wrapped and stored in the refrigerator or thrown out.

▶ Potentially hazardous foods can be safely held at room temperature for no more than a total of two hours. For example, if they've been set out for one hour and then refrigerated, they can safely sit out for one more hour. If these foods have been held at room temperature for more than an accumulated period of two hours, they should be thrown out.

Safe Cooking
170°F (77°C)

Hot Soup
140° F (60°C)

Refrigerator
40°F (4°C)

Water Begins to Freeze
32°F (0°C)

Frozen Food
0°F (-18°C)

Picnics and Bag/Box Lunches

Pathogens can grow to dangerous levels when potentially hazardous foods are held at room temperature for two hours or more. We need to provide some way to protect these foods when packing lunches or picnics that will be eaten several hours later.

When we pack lunches for school or for work, we may make them early in the morning, or the night before. They are often held in the danger zone temperature range for five or more hours. Because we want to provide nutritious lunches, we may include foods like sandwich meats, eggs or milk products. These are potentially hazardous foods.

Perhaps the sickness that children "pick-up" so often at school doesn't always come from being exposed to germs from other children—they could be reacting to pathogens growing in their lunches. Hot or cold thermoses and insulated lunch packs are available to make it easier to keep hot foods hot, cold foods cold and keep lunches safe for our families. Portable ice coolers can keep potentially hazardous foods cold for safe picnics.

Unless you have a dependable way of keeping the foods hot or cold, choose less hazardous foods when packing a lunch or picnic. The following foods can be left at room temperature for up to six hours:

- nuts and peanut butter
- bread and crackers
- cookies and cake
- butter, margarine, and cooking oil
- dry cereal, powdered milk (until reconstituted)
- raw, cooked, or dried fruit
- yogurt (that has had nothing, such as fresh fruit or granola, added by you)
- raw vegetables
- pickles and condiments such as ketchup, relish, or mustard
- dry or hard cheeses like Parmesan (but not cheddar)
- sandwich meats such as salami and pepperoni (but not hotdogs or bologna)
- canned foods (until opened)
- fruit pies

Preparation

▶ Use insulated lunch boxes or picnic baskets to keep any potentially hazardous foods as cold as possible.

▶ Small freezer packs that fit easily into a lunch box are now available in many grocery, hardware and camping supply stores.

▶ Larger freezer packs or block ice can be used to keep the picnic basket or chest cooler cold enough to hold foods safely. Coolers provide much better insulation than baskets and will keep foods safe longer.

▶ Check to ensure that your child's day-care center has refrigeration space for storing lunches.

▶ You could freeze sandwiches (without the tomato and mayonnaise), cooked chicken pieces or pizza slices overnight. They will thaw in about three hours and keep the rest of the foods in a lunch box or picnic basket chilled until lunch time.

▶ Individual-sized boxes of juice can also be frozen and packed with the lunch. They will thaw to a "slush" state by lunch time and help to keep other foods around them cool.

▶ Any foods packed in a lunch box for an afternoon snack should be chosen from the non-hazardous group of foods such as dried, very sweet, very salty, raw fruits and raw vegetables—all properly washed, of course.

Camping

There are many different definitions of "going camping." They can range from heading back into the bush with nothing but a backpack and a bed roll; to setting up a base camp complete with tents, camp stoves and coolers; to travel trailers with electrical or gas powered appliances.

No matter what kind of camping you enjoy, there are several things we need to consider to make certain we keep our food safe under these conditions.

Drinking Water

Finding clean, safe, water can be a concern for campers. If treated water is not readily available, you may need to boil water to ensure it is safe for drinking, cooking and washing dishes.

Pathogens—including bacteria, viruses and parasites—can be carried in water and can make us ill if we drink it or use it for cooking or even when brushing our teeth. This can happen in remote, seemingly pristine areas as well as in populated, industrialized places where facilities for clean water are not available or not working.

▶ You can destroy bacteria, viruses and parasites by bringing the water to a rolling boil for two minutes. It should be stored with a cover to prevent any new pathogens from being introduced through general contamination from the air, dust, sneezes or insects.

▶ Chlorine tablets for purifying water are available from camping supply stores. In order to be effective, chlorine needs to be used in the proper ratio and for the correct amount of time. Follow directions carefully.

▶ Chlorine will destroy bacteria and many pathogens, but it will not destroy *Giardia* and other parasites. If there are reports of parasites in the water in the area where you are camping, boil the water to be certain it's safe.

▶ If you are using liquid chlorine bleach to purify water, the recommended ratio is four drops per gallon of water. Let it sit for 20 minutes to ensure that bacteria and viruses are destroyed. Remember that chlorine will not destroy parasites.

► Another chemical that can be used to purify water is iodine, which is available in tablet, crystal and liquid forms. The manufacturer's instructions will describe the most effective way to use the product.

► Pregnant women, people with thyroid conditions or people with allergies to iodine should consult a doctor before drinking water purified with iodine.

► If you are using flavor crystals to mask the taste of the chlorine or iodine, you must wait until after the recommended standing time has passed to add them.

Dried or Canned Foods

Protecting dried or canned foods under some camping conditions can be challenging. These foods must be kept dry. They should be placed in containers, when possible, to avoid damage to the packages or cans.

If you are back-packing, choose foods that do not require refrigeration such as dried milk powder, dried soups or the new "retort pouch" packaged entrées. Remember, once you add water or open canned or retort pouch foods, they become potentially hazardous and must be protected by cold temperatures of 40°F (4°C) or colder.

These foods can only remain in the danger zone temperature range for two hours and still be safe to eat. So it is recommended that you pay close attention to portion size of servings and only make up what you will use for each meal.

Portable Ice Coolers

Protecting potentially hazardous foods can be one of the biggest challenges for campers. If they are used properly, portable ice coolers make it possible to keep these foods cold. A dependable supply of ice or reusable freezer packs can make the difference between a safe camping trip and a very unpleasant holiday.

► Make sure the cooler is in good condition and that the lid seals tightly when closed. A cold temperature thermometer will let you know how effectively your cooler is actually working.

► To prevent cross-contamination of ready-to-eat foods from meat drippings, use a separate cooler for raw meats and another cooler for ready-to-eat foods.

► You could also have separate coolers for items that will be used frequently. Each time the cooler is opened, air will get inside warming the food and melting the ice faster.

► If using separate coolers is not possible, store items that will be used frequently in the tray at the top or where you can get to them quickly.

► Keep foods wrapped or in water-proof containers. Zip-lock plastic storage bags or sealable plastic containers can help to protect the foods from damage by the water produced by melting ice as well as cross-contamination in the water from other foods or people's hands.

► Since most hazardous foods have a three day storage life at refrigeration temperatures of 40°F (4°C), you may want to bring along some frozen foods. They will keep longer and help keep the cooler cold.

► Smaller items could be used in the first few days and larger items could be used as they thaw. It is important to make sure that drippings from thawing meat don't contaminate anything inside the cooler including the water from melting ice.

Water Begins to Freeze
32°F (0°C)

— Refrigerator
40°F (4°C)

— Frozen Food
0°F (-18°C)

Ice

► Blocks of ice last longer than cubed or crushed ice.

► The colder it is inside the cooler, the longer the ice will last. If you use ice in the cooler with the frozen foods, it can last several days depending on the amount of frozen foods. If you replace blocks of ice in the frozen food cooler and rotate this ice to the potentially hazardous food cooler, you may be able to extend the life of your ice somewhat.

► Purified water frozen in sanitized plastic containers could be used for drinking water as it thaws.

Dry Ice

► Dry ice is made of compressed carbon dioxide gas. Frozen at -109°F (-78°C), it can keep foods or conventional ice frozen solid. If you are camping in a remote area, this could be very beneficial. Dry ice is only available in major centers. Make sure you ask the dealer about the proper handling procedures.

► Depending on weather temperatures, a 15-pound block will last about three days. It is so cold it will freeze anything in the cooler.

► Dry ice will be most effective if you place it in the top center of the cooler so the cold can cascade down around the food. The tray must have holes in the bottom for this to work well.

► You will need a separate cooler with conventional ice for the foods you want to use first.

► You can increase the number of cold temperature holding days by storing blocks of conventional ice in the frozen (dry ice) cooler.

► Make sure to follow the manufacturer's instructions and DO NOT TOUCH dry ice because it can cause serious burns to your skin. Avoid breathing the carbon dioxide vapors.

► Do not let dry ice touch the mouth or use it in drinks. Keep a close eye on children and keep them from touching it.

Cleaning Camp Dishes

► All of the principles of correct dish washing apply, even in a camping situation. It may be necessary to utilize large pots or to empty and refill your wash basin for some of the steps of washing, rinsing, and sanitizing.

► Scrape away as much food as possible. Rinse away any food residue.

► Fill your wash basin with dish soap and hot water, about 110°F (44°C). Wash the dishes, preferably with a plastic bristled dish brush. They are much easier to sanitize than dish cloths.

▶ Place the clean dishes in a second basin of clean, hot water to rinse them.

▶ Place the rinsed dishes in a third basin of hot water with one ounce (30 ml) of bleach per gallon (4 L) of water for a few minutes to sanitize them. If you don't have a third basin, you could empty and rinse your wash basin and use it to hold your sanitizing water.

▶ Sanitize your dish brush or cloth each time you sanitize your dishes. You could use the leftover sanitizing water to sanitize the table cloth or your food preparation area.

▶ Air dry your dishes if they can be protected from general contamination from dust, sneezes or insects while they dry.

▶ If you choose to towel dry dishes so they can be stored, use disposable paper towels. If you must use cloth towels, use a fresh one every day and let it air dry after each use. Don't let people use the dish towel to dry their hands. It should be used for dishes only.

▶ Store your dishes and utensils where they will not become re-contaminated.

Hand Washing

Keeping one's hands clean where there is no hot and cold running water can be very difficult. But, that doesn't make it any less important when it comes to handling food.

If you keep in mind that people—especially the food handler— are the most common source of contamination, it makes sense that hand washing is always necessary.

During the course of preparing the meal your hands may contact raw foods such as the fish you just caught, which we can expect to be contaminated. If you do not wash your hands after touching the raw food, you can pass that contamination to other ready-to-eat foods, dishes or cooking tools.

▶ Wash your hands with soap, rinse and dry them, before you begin to prepare the meal. Even cold water is better than none, but if you can, use hot water.

▶ Keep a wash basin and rinse water handy so you can clean your hands again after you have touched raw foods and before you handle anything else.

▶ Single-use or disposable paper towels are recommended for drying hands because cloth towels can carry the contamination. Every time you use the same cloth towel your hands could become re-contaminated

▶ Change the wash and rinse water when it gets dirty. Clean water is the most important element in effective hand washing because pathogens are not destroyed by soap or by water temperatures that our hands can tolerate.

People at High Risk

People who are very young, elderly or ill face a greater risk from foodborne illness because their immune systems are immature, waning or injured.

A youthful or middle-aged person who is healthy may be able to combat the pathogen *Salmonella* or the toxin produced by *Staphylococcus* and successfully recover after a bout of nausea, vomiting and other discomforts. On the other hand, a person who falls into the "high risk" group, may be exposed to the same pathogen or toxin, and because their immune system is not able to defend as strongly, they can suffer severe illness, slower recovery, or in the worst case, death.

If someone in your household is at high risk to foodborne illness, **all of the precautions** and food handling principles described in this book **should be followed diligently.**

In addition, there are several precautions that can be taken.

► Eggs should be thoroughly cooked. It takes approximately four minutes to destroy *Salmonella* in a fried egg and seven minutes to make sure a boiled egg is cooked thoroughly. The yolk and white should be firm, not runny.

► Use only pasteurized eggs or egg substitutes in recipes that do not call for thoroughly cooked eggs, such as Caesar salad dressing.

Eating raw eggs is not recommended

► Many municipal water supplies are treated to control pathogens such as bacteria and viruses, but many do not use the filters that would be required to remove parasites.

► You can destroy bacteria, viruses and parasites by bringing the water to a rolling boil for two minutes. It should be stored with a cover to prevent any new pathogens from being introduced through general contamination from the air, dust, sneezes or insects.

- If you are buying bottled water, make sure to read the label. Some brands of bottled water are purified while many others are not treated and could be contaminated.

- Pay very close attention to the time and temperature rules and keep potentially hazardous foods out of the danger zone temperature range whenever possible. If a potentially hazardous food has been left in the danger zone temperature range for more two hours, it should not be eaten.

- Do not give honey to infants under two years old. Cases of "infant botulism" have developed when *Botulinum* bacteria grew in the stomachs of babies. The bacteria were able to vegetate from spores introduced in the honey. Pasteurized honey is not safer than raw honey because the pasteurization process does not destroy the spores.

- Pay close attention to "best before" dates. If a person requires reading glasses to see small print, they should be worn when checking expiry dates.

- Make sure to use leftovers within two or three days. Seafoods, casseroles, soups, stews or any food with many ingredients that are handled frequently should be used within two days.

- Do not eat raw seafoods such as sushi, ceviché, mussels, clams or oysters.

Eating raw mollusks is not recommended

- Do not eat uncooked meat. Even rare or medium-rare meats with an internal temperature of 145°F (63°C) or less could contain bacteria that survive cooking because they do not reach temperatures hot enough to destroy them. In well done meats the pathogens are destroyed when the internal temperature has reached 170°F (77°C).

- Choose restaurants that pay attention to good food handling practices, when dining out. If cooked food is not served "piping hot" it should be returned because it could contain pathogens.

Eating uncooked meat is not recommended

KITCHEN MANAGEMENT

Shopping

Shopping or hunting and gathering food is the most basic daily activity of humankind. Of course, most modern hunters find themselves in grocery stores and markets looking for good quality foods at reasonable prices.

We can buy food products from almost anywhere in the world. Foods are available to us in a wide range of forms such as fresh, frozen, dried, pre-mixed, processed or ready-to-eat. They can be packaged in plastic containers, plastic wrap, cardboard, tins, jars, vacuum packs, foil pouches, or atmospheric packs.

The retail food industry faces and meets many challenges in the delivery of good, safe, food for consumers. It seems that new developments and innovations in products and packaging are constantly on the shelf or on the horizon.

In most countries, protection of retail foods is regulated by government agencies. With so many products coming from so many countries in so many forms, these agencies must work hard to ensure that consumers are provided with safe food products.

There are many things to consider when you shop for food. Details for purchasing specific foods are included in each section of this book where a particular food is discussed—for example "Purchasing Finfish". There are some general rules that you can apply when shopping for food.

► Always buy from a reputable source. Food that has been inspected or is processed under government regulations is labeled with a unique code so that if there is a safety concern, it can be recalled by the lot number. If you do not know where a product came from, you will have no way of knowing if it is from the problem lot.

► Don't buy food from an unknown supplier. If you buy products directly from a producer such as a farmer or a fisherman, choose someone who has a good reputation in the community.

▶ Check the store or market for cleanliness. Dirty floors, shelves, refrigerators, rotting foods or unclean food preparation areas are indicators that a high number of microbes are in the area. The more opportunity there is for these organisms to grow, means more opportunity for food to become contaminated.

▶ Check the temperatures in display and storage coolers. Foods that require refrigeration must be stored at 40°F (4°C) or colder, at all times.

▶ Frozen foods should be frozen solid. Even open frozen food display areas must keep frozen foods frozen.

▶ Displays that contain both raw and ready-to-eat foods must be set up in such a way that there is no chance for cross-contamination to occur. For example, in a seafood display, ready-to-eat foods should be displayed in a container that will protect them from the display ice which can easily become contaminated with drippings from the raw seafood. The ready-to-eat food should also be located so that drippings from raw foods will not fall on them when raw foods are taken from the display.

▶ Read labels to make sure that the food has been stored properly by the grocer. New packaging methods require new storage conditions. Don't assume that because the food is in a tin or a vacuum package, it is safe at room temperature. Always assume that the manufacturer knows how the food must be stored and has labeled it appropriately.

▶ Check "best before" dates. Only buy foods that you intend to use before that date. A great sale price is not necessarily a bargain if the food has been stored longer than the manufacturer recommends. If you find newer products stored at the front of a shelf or refrigerator it can be an indication that the people who stock the shelves are not following the "first in, first out" rule.

▶ Protect ready-to-eat foods from contamination by raw foods while you are shopping. For example, if a package of uncooked meat is leaking, place it inside a plastic bag to keep it from dripping onto lettuce, or other ready-to-eat foods or food packages.

▶ Ask the person bagging your groceries to keep potentially hazardous foods such as uncooked meats, separate from other foods.

▶ Purchase foods that require refrigeration or freezer storage at the end of your grocery shopping trip so exposure in the danger zone temperature range is minimized.

▶ Make grocery shopping your last errand so that perishable foods can be protected in the refrigerator or freezer as soon as possible.

Storing Foods

Refrigerated Storage

▶ Thermometers for measuring the temperature in your refrigerator are the best way to determine if it is actually cold enough to protect your food. Locate the thermometer at the top of the compartment near the door, which is usually the warmest spot. The recommended temperature for holding cold food is 40°F (4°C) or colder.

▶ Quickly cool hot foods before you put them in the refrigerator. If you place hot food in the refrigerator, the temperature inside the refrigerator rises and all of the food could then be in the danger zone temperature range.

▶ Good air circulation is important for the even distribution of cold air throughout the refrigerator. Separate items on the shelves, and don't create warm spots by placing warm or room temperature food close together.

▶ All foods should be wrapped or stored in sealed containers to prevent them from becoming contaminated.

▶ As an extra precaution, raw food should be well sealed or wrapped and stored on a shelf below ready-to-eat foods. This will prevent the drippings of raw foods from contaminating other foods.

▶ It is a good idea to label and date containers. Most prepared foods and many canned foods need to be used within three days of cooking or opening.

Water Begins to Freeze
32°F (0°C)

Refrigerator
40°F (4°C)

Frozen Food
0°F (-18°C)

Frozen Storage

▶ The freezer compartment in refrigerators, including the ones with a separate door, are not designed to keep foods as cold as chest freezers. You can expect foods stored in the refrigerator freezer to have a shorter storage life than foods stored in a chest freezer.

► The thermometers for measuring the temperature in your refrigerator can also register the temperature in your chest freezer. Locate the thermometer near the door, which is usually the warmest spot. The recommended temperature for holding frozen food is 0°F (-18°C).

► Air in a package of frozen food will reduce the storage life of the food. When packaging food for freezer storage, remove as much air as possible from the wrapping or freezer bag.

► All foods should be labeled with the content and date so that you can use the oldest foods first.

► There will be less wear and tear on your chest freezer if prepared foods are cooled in the refrigerator before they are put in the freezer. If you place warm food in the freezer, the temperature inside will increase and any nearby food could be warmed.

► Good air circulation is important for the even distribution of cold air throughout the freezer. Separate items and don't create warm spots by placing unfrozen food close together. Freezer racks that allow air circulation should be used while the food is freezing.

► Avoid overloading the freezer. Restrict the load to the amount of food that will freeze solid in 24 hours.

Power Outage Concerns

If your electrical power is interrupted for a short time, you can substantially reduce the loss of cold by keeping the lid of your chest freezer or the door of your freezer compartment or refrigerator closed as much as possible.

► A chest or upright freezer can keep your food frozen for one to two days depending on the room temperature and how full the freezer is. Food in a full freezer will stay cold longer than food in a partially full one.

► Depending on how full it is and the temperature of the room, the freezer compartment on the top of your fridge will keep food frozen for four to six hours.

▶ If dry ice is available, you could use it in your freezer. Make sure to follow the manufacturer's instructions and DO NOT TOUCH it because it can cause serious burns to your skin. Avoid breathing the carbon dioxide vapors.

▶ Unless there are signs of discoloration or an "off" odor, food that has thawed to the point where it still contains ice crystals, can be refrozen.

▶ If food has been in the danger zone temperature range for more than two hours it should be thrown out.

Dry Storage

▶ Dry goods such as grains, cereals, and legumes, as well as tinned and glass container foods must be kept dry.

▶ Good air circulation is important to prevent the build up of moisture. It is best to store these foods in an area where they will not be exposed to moisture from dish washing, cooking, or steaming kettles.

▶ Dry goods are best stored at cool-room temperatures of 32° - 50°F (0° - 11°C). They can also be stored safely at normal room temperatures, but they may not keep as long as they would in a cool-room. Kitchen cupboards with the doors closed are generally cooler inside than the rest of the kitchen.

▶ All dry foods should be in sealed containers to protect against contamination from dust, insects and rodents.

▶ Dry goods should be protected from direct sunlight. Light can cause photodegradation by oxidizing the fats in the food. Most people will not eat foods that have a rancid flavor but some people do not seem to be affected by the change in flavor. Oxidized fats are thought to be carcinogenic and mutagenic.

▶ It is a good idea to label and date any containers that you transfer foods into. Then you will have a way of determining how long an item has been in storage. Certain dry foods, like whole wheat flour, have a relatively short storage time before the enzymes begin to oxidize fats.

▶ Rotate the foods in your cupboard by placing the newer ones at the back and pulling the the older ones forward to be used first.

Personal Hygiene

The food handler has been found to be the source of contamination in the majority of foodborne illness cases. If you are the person cooking, then this means that you should pay close attention to your personal hygiene.

Our bodies are normally covered with *Staphylococcus* bacteria. Because we touch food with our hands during preparation, our hands can be a source of cross-contamination if we handle raw foods such as ground beef that contained *E. Coli* bacteria.

This is also why it is recommended that we use clean utensils whenever possible rather than our hands, when we prepare food. If we reduce hand contact, we reduce the chance of transmitting pathogens to the food.

Learning and practicing good personal hygiene is a very important step in the safe preparation of food. The simple practice of frequent hand washing before and during food preparation can significantly reduce the risk of food contamination.

Infected Cuts

Staphylococcus bacteria, which are responsible for numerous cases of the "24 hour flu" or foodborne illness live harmlessly on skin, in noses and in throats of many people around the world. They usually don't cause any serious problems to us—unless they are allowed to grow in our food. These bacteria are often found in high numbers in infected cuts, burns, pimples or boils. If a person has an infected cut on a hand or finger, they can then transmit high numbers of bacteria to any food they touch. Because bacteria multiply so rapidly, each added bacteria can reproduce millions like itself.

Since bacteria are microscopic, they can protect themselves deep inside a cut. Hand washing alone, is not enough to make it safe to touch food while the infection exists.

A band-aid doesn't provide enough protection for handling food. As soon as it becomes moist, the bacteria can multiply in the band-aid itself. Then the band-aid becomes a source of many more pathogens.

You could cover the cut with a waterproof protector or wear a latex glove to prevent the bacteria from spreading to the food. Gloves require special attention to make sure they won't transmit pathogens.

Latex Glove Use

Latex gloves can provide a waterproof barrier between the pathogens on a person's hands and food. Unless someone in the household must be protected from foodborne illness due to a weakened immune system, most people will not find it necessary to use these gloves in the home setting.

If you want to protect food from an infected cut on a hand, you may chose to wear a glove on that hand. One of the chief concerns with the use of gloves is that the wearer can often forget that even though their hands are clean, the glove itself can become contaminated. Because you cannot feel your hands become soiled when they are inside gloves, you are less likely to wash them frequently. It is recommended that you only wear one glove so you can "feel" when your hands need to be washed.

Gloved hands need to be washed for all of the same reasons that bare hands do. Gloves should be changed after every chance of contamination—such as handling raw meat or taking out the garbage.

Illness

When a person is ill, the body's immune system is overwhelmed by pathogens. Preparing food for other people while you are ill could pass the illness on to them. Good hand washing practices alone may not be enough to prevent contamination of food, because the sheer numbers are just too high. If possible, have someone else take over cooking responsibilities until you recover.

Because any pathogens from the mouth, throat, or nose can be transported to the food, it should be protected from coughs or sneezes. If you must cough or sneeze, cover your mouth with your hands to prevent pathogens from being projected onto food or work surfaces. If a tissue is available, it can help capture the cough or sneeze. Always wash your hands afterwards.

Hair

Hair is routinely contaminated with microbes. If a hair falls into food it could transport pathogens with it. The average person loses approximately 80 hairs each day. If you are preparing food for someone who is ill or for a large gathering, you may want to take the extra precaution of securing your hair.

Hand Washing

The importance of good hand-washing practices in preventing the spread of disease has been emphasized for many years.

If you keep in mind that people are the most common source of contamination, it makes sense that frequent hand washing is always necessary. During the course of preparing a meal, your hands may contact raw foods which we can expect to be contaminated. If you do not wash your hands after touching the raw food, you can pass that contamination to other ready-to-eat foods, dishes, counter-tops or cooking tools.

► Jewelry should be removed before washing your hands to prepare food. It is difficult to clean around watches, bracelets and rings. If jewelry becomes moist, bacteria can grow in the setting.

► Soap is important for breaking down oils and soil on the skin so dirt and pathogens can be removed.

► Running water works best to rinse the soap, dirt and pathogens from your hands. If running water is not available the best alternative is a basin of water that is changed frequently.

► It takes about 30 seconds of brisk rubbing with soap and water to lift the dirt and pathogens out of your pores. Rub your hands together in a rotary motion to create friction.

► A fingernail brush should be used to clean around the fingernail beds and under the fingernails themselves. Some people have very deep lines in their hands and so they may need to use a fingernail brush to clean them.

► Hand-drying towels should never be used for wiping dishes or counters. Cloth towels are a breeding ground for bacteria. Every time you use that cloth towel, your hands could become re-contaminated.

► To reduce the risk of cross-contamination from hands to dishes, use separate towels for drying hands and dishes. Both towels should be replaced with clean ones each day.

► Because cloth towels can carry contamination, disposable paper towels are recommended for drying hands in high-risk situations like camping.

Wash your hands:
► **Before handling any food; to prevent transmission in the first place.**

► **After handling raw food; we must expect these foods to be contaminated.**

► **After sneezing or coughing into your hands; pathogens are often found in the nose, throat or mouth.**

► **After smoking; fingers touch the lips and become contaminated with saliva.**

► **After toilet use; the most serious pathogens are often transmitted through the "fecal-oral" route. When someone who is infected does not wash their hands after using the toilet anything they touch, especially food, can transmit the disease.**

Water

Drinking Water

Clean, safe, water is one of our most vital needs. Because it has such an important impact on the health of any population, the safety of drinking water in most developed countries falls under government regulation and responsibility. In most public water systems, water is treated with chemicals such as chlorine to destroy pathogens. It may also be run through filters as an extra precaution to remove bacteria and parasites.

Each year, throughout the world, human illness has resulted from drinking contaminated water. Pathogens, including bacteria, viruses and parasites, can be carried in water. These microbes can make us ill if we drink contaminated water, use it for cooking or even for brushing our teeth. This can happen in remote, untouched areas as well as in populated, industrialized places where facilities for clean water are not available or not working.

The safety of tap water can vary from country to country as well as from community to community. If you are travelling it is important to know if the water in that community is safe to drink directly from the tap or if it requires treatment to make it safe.

If treated water is not readily available, you may need to purify water to ensure it is safe for drinking, cooking and washing dishes.

▶ You can destroy the pathogens by bringing the water to a rolling boil for two minutes. It should be stored with a cover to prevent any new pathogens from being introduced through general contamination from the air, dust, sneezes or insects.

▶ Chlorine tablets for purifying water are available from camping supply stores. Chlorine needs to be used in the proper ratio and for the correct amount of time to be effective. Follow the directions carefully.

▶ Chlorine will destroy bacteria and many pathogens, but it is not effective against *Giardia* and other parasites. If you hear reports of parasites in the water in the area where you are, the water should be boiled to be certain it is safe.

- If you are using chlorine to purify water, the recommended ratio is 4 drops per gallon of water. Let it sit for 20 minutes to ensure that bacteria and viruses are destroyed. Remember that liquid chlorine bleach will not destroy some parasites.
- The chlorine taste and odor will dissipate if the water is stored in the refrigerator for a few hours.
- Another chemical that can be used to purify water is iodine which is available in tablet, crystal and liquid forms. The manufacturer's instructions will describe the most effective way to use the product.
- Pregnant women, people with thyroid conditions or allergies to iodine should consult a doctor before drinking water purified with iodine.
- If you use flavor crystals to mask the taste of chlorine or iodine, you must wait until after the recommended standing time has passed to add them.

Water Filters

Many people now use filters in an effort to remove tastes, odors, solids, chemicals or micro-organisms from their drinking water.

There are several types of water filters available. Different models use different materials or a combination of materials in the filtering elements. Units that are intended to remove certain tastes or odors from water may not be designed to remove pathogens.

Some models are designed to remove smaller bacteria, others only remove larger bacteria, parasites and helminths. The manufacturer's instructions will give directions for cleaning the unit and indicate when filters should be replaced.

- Follow the manufacturer's instructions carefully. Failure to clean the unit or change the filters could result in a build-up of pathogens inside the unit.
- Some filters are cleaned by abrasion. As the filter walls gets thinner the opening grows and larger pathogens may get past the filter as it wears.
- The effectiveness of the filter can be affected by the amount of contaminants in the water, and the life of a filter may be shorter where the water is heavily contaminated.

Two-part Water Systems

Two-part water systems use separate lines to carry untreated water and drinking water. One part of the system carries water for laundry, bathing, and watering plants. The other part is used to carry potable water used for drinking, cooking and dish washing.

Since safe drinking water is so essential to good health, it is important to ensure that these types of dual water systems be kept separate and that cross-contamination from one system to the other is not allowed to happen.

Kitchen Equipment

A well-equipped kitchen may have many appliances and the best materials for counter tops and floors. But, some of the things that will really make a kitchen work to provide safe food for our families may actually be rather small and inexpensive.

A FOODSAFE tool kit: including a one-gallon pail, household chlorine bleach, a plastic spray bottle, a clean cloth and a probe thermometer can help you make sure the food you serve is safe.

Other easily affordable items include:

▶ Masking tape or labels and permanent felt markers for dating foods going into the refrigerator or freezer.

▶ Plastic, gel-filled cooling wand stored in the freezer until needed for cooling hot foods quickly.

▶ Plastic, gel-filled ice packs for cooling lunches and picnic coolers.

▶ Quality thermoses that will keep hot foods hot or cold foods cold for lunches and picnics.

▶ Insulated lunch boxes or bags.

▶ Thermometers to measure refrigerator or freezer temperatures.

▶ Re-sealable plastic containers that can be sanitized for protecting foods in storage.

▶ Plastic wrap, plastic bags or foil for protecting foods in the refrigerator, lunch box or chest cooler.

▶ Freezer bags or wrap for storing frozen foods.

▶ Plastic bristled brushes used only for cleaning dishes.

▶ Plastic bristled vegetable brushes used only for cleaning raw vegetables or fruit.

▶ Enough dish towels or wiping cloths that they can be changed at least once each day.

▶ Rubber mesh matting for cupboards where glasses and cups are stored upside down.

▶ Hand soap.

▶ Disposable paper towels for cleaning or drying hands.

▶ Separate cutting boards for raw and ready-to-eat foods.

Appliances

The price of an appliance may not be the best indicator of its quality. When we look at these tools with an eye toward food safety, there may be more important things to consider.

Refrigerators

In terms of food safety, refrigerators are second only to clean, hot running water when it comes to tools that have improved the nature of our kitchens. The ability to conveniently hold perishable foods at cold or freezing temperatures means that we can safely keep a wide variety and a large quantity of foods on hand.

Like any other tool, refrigerators need to be in good working order and used properly to do their job effectively.

► There should be good seals around the refrigerator door.

► A completely separate door for the freezer compartment means that foods stored in the freezer are not exposed to warmer temperatures each time the refrigerator is opened.

► Racks in the doors give us quick access to items that are used frequently, reducing the time food in the fridge or freezer is exposed to warmer temperatures while we look for things.

► The thermostat should be set to keep the food at 40°F (4°C) or colder. Even if the food in your refrigerator feels cold to touch, you will not know if it's cold enough unless you test with a thermometer that is designed to register cold temperatures. Take the reading at the top of the refrigerator compartment, close to the door, which is usually the warmest area.

► Special meat trays or compartments are designed to keep meats just above freezing temperature of 32°F (0°C).

► Crispers or produce drawers are designed to prevent produce that has a high water content, such as lettuce, fresh herbs, or mushrooms from freezing if the

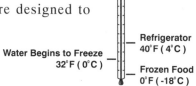

Water Begins to Freeze
32°F (0°C)

Refrigerator
40°F (4°C)

Frozen Food
0°F (-18°C)

temperature in your refrigerator approaches 32°F (0°C). Moisture loss is also reduced so our produce stores better in these drawers.

▶ Storing eggs in a built-in egg rack in the side of the door is no longer recommended. Eggs may develop hairline cracks when the door is opened and closed and the cracks could allow contamination inside. If there is *Salmonella* on the surface, it could cross-contaminate other foods in the refrigerator. Store eggs in their original container.

▶ Some refrigerators built before the mid-1960's had shelves made with cadmium. It was discovered that if acidic foods came into direct contact with the shelf, they could become contaminated by the metal and cause food poisoning. Cadmium shelves are silver in color, but have a flat or matt surface rather than a shiny one. Shelves from old refrigerators should never be re-used for home-made barbecue grates.

▶ It is recommended that you clean your refrigerator regularly and that it be washed, rinsed and sanitized often. It is important to prevent a build-up of pathogens on the walls, shelves or in the drawers. Refrigeration temperature does not stop bacteria from growing; it just slows them down.

▶ One of the most important considerations, when choosing a refrigerator, is cleanability. All racks, drawers, sliding doors should be easy to remove so you can clean them and reach all interior surfaces.

Range Tops and Ovens
▶ Whether you are cooking with electricity, natural gas or propane, the most important feature of these tools is your ability to control the temperature.

▶ The thermostats and controls on cooking equipment should allow you to provide steady heat to foods you are cooking. For example, if the temperature drops sharply and then rises quickly, it can be more difficult to keep a soup simmering at a constant safe temperature. There's the danger the food will drop into the danger zone temperature range during the cooking time.

► It is recommended that you clean your range top after every use. It should be washed, rinsed and sanitized just like any other tool you use to prepare your meal. It is easy to contaminate the surface with pathogens from raw food before and during cooking.

► When choosing a range top or oven, look for easy cleaning features. The element housings or burners should lift out so you can clean them and reach the area underneath them. Some range tops lift up to make cleaning spills easier.

► Microwaves can deliver uneven temperatures to different areas inside the foods. If microwaves come equipped with a turntable, the food will be rotated through the different areas and will be heated more evenly. A good defrost cycle can be a real asset to the busy cook. The defrost cycle should thaw the food evenly. It should not partially cook one area while leaving another area frozen.

► Microwave ovens should be cleaned and sanitized after each use. Pay careful attention to the upper panel, so that the remains of previous splashes are not carried down into the new foods.

Mixers, Blenders and Utensils

These kitchen tools come in many sizes, shapes and styles. Often "fashion" seems more important than function and all too often, cleanability is not even considered.

Blenders used in care facilities have been associated with outbreaks of Salmonellosis. Because they were difficult and dangerous to clean, they were cleaned ineffectively, and the *Salmonella* bacteria survived between uses.

► Choose materials that will stand up to vigorous cleaning and sanitizing chemicals or heat.

► Check that you can disassemble the item and effectively clean all parts.

► Smooth, tight seams, rounded angles rather than sharp corners and hard surfaces will allow easier cleaning and prevent food and pathogens from collecting on these tools.

► It is important to prevent any build-up or "crusting" of food residue or grease. Pathogens can survive and find shelter in this material and spores can remain hidden, ready to vegetate and grow as soon as moisture is available.

► Wooden spoons should be in good condition with smooth surfaces. It may be necessary to sand them if they become scratched or tattered. If there are any deep scratches or if the wood begin to split, they should be replaced. Wooden spoons should be cleaned and sanitized and allowed to thoroughly dry between uses.

► Plastic or rubber spatulas should be in good condition. If deep scratches develop or if the rubber begins to crack they should be replaced. It is no longer possible to effectively clean them.

Cutting Boards

In many cases of foodborne illness, a contaminated cutting board has been found to be the source of the pathogenic microbes.

For example, if raw pork was prepared on the cutting board, we can expect the cutting board to be contaminated with *Campylobacter* or other bacteria. If the next food we prepare on that board is lettuce, we can expect the lettuce to become contaminated with the *Campylobacter* bacteria on the board. Since we will not cook the lettuce, or any ready-to-eat ingredient, we will not destroy the bacteria. When we eat the lettuce we can become infected with *Campylobacter* bacteria.

If we keep in mind that microbes are very small, we can see that what looks like a tiny scratch on the cutting board to us will be like the Grand Canyon to them. This is why washing, rinsing and sanitizing cutting boards is so important.

► Cutting boards can be made of food-grade wood, hard plastic, marble or glass.

► They should be washed, rinsed and sanitized after each different use.

► If you use a scrub brush, you will be able to clean inside any scratches on the surface of the cutting board. You need to remove all organic material so the sanitizer can make direct contact with all surfaces to destroy the microbes.

► Wooden cutting boards should be planed or sanded regularly to remove scratches.

► The harder the plastic in a cutting board is, the less likely it is that it will scratch easily. Once a plastic cutting board becomes badly scratched, it will be very difficult to effectively clean and sanitize. It is best to replace it.

► It is recommended that you have separate cutting boards for raw and ready-to-eat foods. This way you will reduce the risk of cross-contamination. Colored cutting boards are now available so you can easily identify which board is for which kind of food.

Dishware, Cutlery, Pots and Pans

When we purchase kitchen essentials like dishware, cutlery, pots and pans, we are often concerned with how they will look on the dining room table or in the kitchen. It also important to choose them based on their cleanability and how well they will stand up to cleaning and sanitizing.

► Choose materials that will stand up to vigorous cleaning and sanitizing chemicals or heat.

► Smooth, tight seams and rounded angles rather than sharp corners on cutlery and pots and pans will allow easier cleaning and prevent food and pathogens from collecting in seams and corners.

► Smooth-surfaced dishes are much easier to clean than dishes with a pattern impressed into them.

► It is important to prevent any build-up or "crusting" of food residue or grease. Pathogens can survive and find shelter in this material and spores can remain, ready to vegetate and grow as soon as moisture is available.

► Check to make sure the glaze on dishware is in good condition. Sometimes there can be flaws which expose the porous material underneath the glaze. It would not be possible to effectively sanitize dishware in this condition.

► Any plates, cups or bowls with cracks in the glaze or chips should be replaced. Pathogens can survive and be protected under any organic material that builds in the crack or chip. Dish washing and sanitizing will not be effective enough to remove the pathogens.

► Glassware should have a smooth, rounded surface inside. If there is a recess in the bottom, it could very difficult to clean.

► Cutlery should have a smooth surface, especially in areas that will touch the mouth, like the bowl of the spoon or tines of the fork. If the cutlery has been formed with a stamp, like some less expensive brands, the area between the tines of the fork may have a rough surface as a result. This surface must be be smooth in order to get it clean. If there is a pattern formed in the handle it should end well before the part that touches the mouth.

► Chopsticks may be made of wood, plastic or have a lacquered finish. They should be replaced if scratches or chips develop on the surface. Wooden chopsticks are more likely to scratch and they are difficult to clean effectively. They should be cleaned and sanitized after each use.

Kitchen Design

Most kitchens are designed to allow for convenient meal preparation. Sometimes trying to "make things work" around existing electrical hook-ups or plumbing can produce less than convenient results. The actual lay-out of work areas and food or dishware storage areas can make a difference to how easy or difficult it is to prevent cross-contamination and other food hazards.

Food Storage Areas

► Dry, canned or glass-contained foods should be stored away from any possible dampness, from steam or from sweating plumbing pipes. Moisture promotes microbe growth in dry foods and on walls or shelves. Moisture causes rust to form on canned goods which can affect the seal.

► Doors on cupboards will keep foods cooler and protect them from general contamination from dust and photodegradation by direct sunlight.

► Cupboards and storage areas should be cleaned and sanitized regularly. Food spills will attract pests like insects and rodents that bring even more pathogens into your kitchen. Of course any pathogens already in your kitchen could find food spills a good place to grow and multiply if conditions are right.

► Always store detergents, cleaners or chemicals in a separate area of your kitchen where they will not come in contact with food or be mistaken for food or food products.

Work Areas

► Counters and work areas should be made of a hard, smooth material. Any seams should be tight and flush to prevent a build-up of grease, grime and microbes. Coving or rounded seams where vertical and horizontal surfaces meet makes cleaning seams and sanitizing these areas much easier.

► The grouting between tiles should be in good condition and covered with a waterproof coating to prevent moisture, food and pathogens from collecting and to make cleaning easier.

► Wooden counters should be washed, rinsed and sanitized after each use. It is recommended that they be scrubbed with a brush to make sure you have cleaned all surfaces, including inside any small scratches. Visible scratches should be sanded until the surface is smooth again.

► It is recommended that you keep separate work areas for preparing raw or ready-to-eat foods. For example, if you have enough room in your kitchen, you could prepare raw foods on the counter closest to the sink and the other foods farther along the counter. This would keep ready-to-eat foods away from any drippings or splatter.

► If you have limited space, you would need to take the time to thoroughly clean and sanitize the work area each time raw food is prepared.

Garbage Storage

► The daily garbage storage area needs to be both convenient and separate from any food or dishware storage. A cover will help to reduce the numbers of pathogens that could be spread around the kitchen.

► Garbage containers with foot-controlled lids prevent contamination of your hands when you open them. If your hands become contaminated from the lid of the container, it is important to wash them well before handling food.

► Garbage should be removed from the kitchen area frequently. Outside storage should be far enough from your kitchen to keep insects and rodents away from food storage areas. Outside storage containers should have tight fitting lids to reduce the number of pests that are attracted to them.

► If you collect waste foods to compost for the garden, they should be removed from the kitchen and the container should be cleaned often. Large numbers of pathogens can be present because compost is the perfect environment for them to thrive.

► If you collect cans, jars or plastic containers to be recycled, they should be cleaned as soon as possible to prevent pathogens and mold from growing on the residue of food inside them. Store recyclables away from food supplies.

Floors

▶ Kitchen floors should be kept in good repair and cleaned regularly to remove dirt and debris from the area.

▶ Kitchen floors are a collection place for microbes. If we remember that they are everywhere and can be carried on dust, moisture and in the air, it is easy to see how gravity can cause them to end up on the kitchen floor. Pathogenic *Listeria* bacteria have been found growing in cracked or broken floor tiles.

▶ It is recommended that kitchen floors be made of a hard, smooth material that will stand up to vigorous cleaning.

▶ To make it easier to clean and sanitize them, seams in kitchen floors should be tight and flush to prevent a build-up of grease, grime and microbes. Coving, or rounded seams where vertical and horizontal surfaces meet also makes cleaning kitchen floors easier.

▶ The grouting between tiles should be in good condition and covered with a waterproof coating to prevent food and pathogens from collecting there and to make cleaning easier.

▶ Food should not be stored directly on the floor. It is recommended that it be stored at least six inches (15 centimeters) off the floor to reduce the possibility of food contamination.

Kitchen Cleanliness

A clean kitchen has long been associated with good food handling practices. Bits of food or drippings from raw meats will attract pests like insects or rodents and spread contamination in the kitchen.

Most people believe that a "squeaky clean" kitchen equals "safe food." What we may not realize is that the pathogens associated with foodborne illness cannot be seen with the naked eye and may be present even in the cleanest looking kitchens. These unseen microbes can be a more serious threat to the safety of our food.

It would be extremely difficult to create a sterile environment (removing all pathogens and spores) in our kitchens and it is not feasible or necessary to do that. Instead, we sanitize them.

Detergents or cleaners remove the organic matter and many microbes from the kitchen surfaces. Sanitizers remove 99 per cent of the microbes but spores may still be there.

If we simply add an extra sanitizing step in our clean-up of kitchen counters, cutting boards and food storage areas, we can reduce our risks of foodborne illness substantially.

NOTE: There are four distinct steps in kitchen sanitization.

1. WASH	3. SANITIZE
2. RINSE	4. AIR DRY

Cleaning

Cleaning includes the washing and rinsing steps of the kitchen sanitization procedure.

Traditional knowledge about "cleaning" products has often been based on removing the dirt and grease that we can see in our kitchens. We may mistakenly believe that because vinegar makes glass and stainless steel shine, it also destroys pathogens. We may think that ammonia smells so bad that it "must" be killing anything it contacts. Alcohol may be effective against some bacteria and protozoa, but it is not effective against a broad range of pathogens.

Some of these products, as well as detergents, do help to remove the organic matter from kitchen surfaces, equipment and cleaning cloths.

Cleaning Procedures

▶ The water used for cleaning kitchen surfaces and equipment must be clean and safe to drink, otherwise, it could contaminate what you are trying to clean.

▶ Remove any crumbs and food debris from the area or item to be cleaned. This may require some water and scrubbing to dissolve dried-on particles. For example, if a cutting board has small scratches on the surface, it may be necessary to use a plastic, bristled brush to make sure you've reached inside the scratches.

▶ Use hot soapy water and a clean cloth to wash the surfaces. Rinse the area or item with clean, hot water.

▶ Rinse your cloth thoroughly with clean, hot water and wring it dry. If it is heavily soiled, put it in your laundry to be washed.

▶ Dishcloths have been a common source of cross-contamination. In scientifically conducted tests, dishcloths taken from a random sampling of homes were found to contain a variety of pathogenic bacteria even when they "looked" clean.

▶ Used dishcloths should be replaced with clean ones at least once each day. Any bacteria present can multiply in the cloth itself, using food or moisture trapped in the fibers of the cloth.

▶ The scrub brush used in cleaning should be sanitized after it is used. This can be done easily when you sanitize your kitchen and equipment.

Sanitizing

There are a variety of "green cleaners" such as borax and baking soda on the market. These products have been promoted as an alternative to traditional sanitizing cleaners such as household chlorine bleach. While alternative cleaners may clean surfaces, it was concluded in scientifically controlled tests, that they were NOT germicidal. They did not sanitize or disinfect. In fact, when alternative cleaners were used on a sponge, the sponge then went on to contaminate other surfaces with the pathogens it had picked up in the initial wiping.

There are several areas of research being developed to produce sanitizers that are more effective for a broad range of pathogens and are less likely to cause environmental damage if they are abused, but in the meanwhile the following products are recommended.

Sodium Hypochlorite (Chlorine Bleach)

The recommended sanitizer for home use is unscented (scented bleaches may leave a sticky film), chlorine bleach. The recommended ratio is one ounce (30 ml) of bleach per gallon (4 L) of water. This ratio is based on a product that has a 5.25 per cent sodium hypochlorite content. Chlorine bleach dissipates quickly in hot water but it takes longer to sanitize at cooler temperatures. Some brands are stronger or weaker and you may want to adjust your ratios accordingly. Used in the correct ratios, this sanitizer is less harmful to the environment and will not disturb the bacterial action in septic tanks when used properly.

Sometimes we can be tempted to conclude that "more must be better." This is not the case with chlorine bleach. The effect of one ounce (30 ml) of bleach per gallon (4 L) of water is quite sufficient for sanitizing equipment and cleaned surfaces in our kitchens. The most you would ever need to use is two ounces (60 ml) of bleach per gallon (4 L) of water and that would only be in extreme situations where you need extra assurance. Higher concentrations are not more effective against pathogens and you would be adding unnecessary amounts of sodium hypochlorite to our environment.

Iodine

Iodine is another sanitizer that is effective against a broad spectrum of pathogens. When used in the recommended ratio of 12.5 parts per million, it will turn water a khaki brown color. Because it can stain cloths and hands, it is not used as commonly as chlorine bleach. Iodine works quickly, even at cool water temperatures.

Other Sanitizers

Other sanitizers are also available, but some of them are only effective against specific pathogens. For example, some will destroy *Salmonella* bacteria but not the *Hepatitis A* virus. The manufacturer's label should give clear directions on their use and effectiveness.

Sanitizing Procedures

The strength of chlorine bleach can be affected by exposure to air, heat or light. Bleach should be stored with the cap tightly closed, in a cool dark area and in a plastic container that you cannot see through. If it is stored properly, bleach has a shelf life of 6 months after purchasing.

- ▶ Fill a bucket or your kitchen sink with one gallon (4 L) of hot water and add one ounce (30 ml) of chlorine bleach.
- ▶ Dip your clean cloth in the bucket or sink and lightly wring, leaving the cloth quite damp.
- ▶ Use the damp cloth to wipe over the area or item you have just washed and rinsed.
- ▶ Allow the area or item to thoroughly air dry before using it again.
- ▶ Rinse your cloth in clean, hot water. Dip your cloth into the chlorine solution and then wring it dry. Allow the cloth to air dry by spreading it out and hanging it rather than storing it crumpled up.
- ▶ You may want to use a spray bottle filled with the chlorine solution for spot cleaning. Because metal can be corroded by the chlorine and the chlorine will lose its sanitizing quality, use a plastic spray bottle for this solution.
- ▶ You still need to thoroughly clean and rinse the area before you spray the sanitizing solution. Don't wipe the area after you spray it. The chlorine will continue to work as the area dries and will evaporate from the surface.
- ▶ Be careful that the spray doesn't get on your face or clothes.
- ▶ The plastic spray bottle should be stored in a dark area such as a kitchen cupboard or you could use an opaque plastic bottle. The solution should be replaced with fresh chlorine bleach and water each day.

Dish Washing

If there is ever a chore that is never finished, it must be dish washing. Hot running water and modern dish washing appliances have made this daily task much easier. But with the added convenience, we may have lost some basic understanding about the principles of "clean" dishes.

In grandmother's kitchen, the dishes were rinsed with a kettle of boiling water. That boiling water was the sanitizing step of old. The introduction of hot, running water, stored in hot water tanks, meant that we could get our dishes clean efficiently. We no longer waited for kettles to boil before the task was complete. But hot tap water is not hot enough to sanitize.

The introduction of mechanical dishwashers has made the never-ending task of dish washing even easier. Not only do we not have to wait for kettles to boil, but we can store the soiled dishes out of sight, in the dishwasher.

In our modern family life, the task of dish washing may not be getting the attention it deserves. Many cases of food borne illness have been associated with improperly washed dishes.

We can still clean our dishes efficiently, with the conveniences we've come to expect. By following a few basic principles of "cleaning" dishes, we can also protect our families from illness.
Remember: there are four distinct steps in proper dish washing: WASH—RINSE—SANITIZE—AIR DRY.

Most families will leave out the sanitizing step in daily dish washing. Unless someone in the family is very young, elderly, or ill (these people are at higher risk to foodborne illness), this is probably not too much of a health risk for the average person.

However, if you want to be on the safe side, or if someone in your family is in the high risk group who are vulnerable to food poisoning, you can add the sanitizing step to your home dish washing procedure. If you host a large gathering of people, it is recommended to add this step in your clean-up because more people can also mean more potential for illness to spread.

Washing Dishes By Hand

Unless the cloth has been sanitized before each washing session, washing dishes with a dishcloth is not recommended. Dishcloths are easily contaminated with pathogenic microbes because they only need microscopic sized food particles to dine on, and dishcloths hold the moisture they need to survive and grow.

Pathogens could grow to high numbers quickly, right in the cloth. Dish soap doesn't affect them and the temperature of the wash water will not be high enough to destroy pathogens. In fact in many cases, it will be close to their favorite temperature of 98.6°F (37°C).

► A dishcloth contaminated with pathogens will allow them to be spread to the dish water and the dishes. If this contamination is on the clean dishes, it can then be spread to any food placed on dishes.

► Plastic scrub brushes are recommended because they can be easily sanitized and air dried rapidly.

► If there is any organic material (food or detergent) left on the dishes, the sanitizer cannot be as effective. The sanitizer will be used up on organic material and won't have a chance to destroy any microbes that may be there. That is why it is important to follow the steps in the recommended order.

► Dishes that are cracked or chipped or that have glaze worn away should be discarded. These dishes are a potential hazard because they cannot be effectively cleaned or sanitized.

Pre-rinse

► Scrape away any chunks of food from the dishes and rinse away as much organic material as possible. The cleaner the wash water is, the cleaner the dishes will be.

Washing

► If you have two sinks you could fill one with hot 110°F (44°C), soapy, wash water and the other with hot, clean, rinse water.

▶ The wash water must be below 140°F (60°C) to prevent "baking on" food particles, while the rinse water can be as hot as you like.

▶ The dish soap helps lift the organic material, including any pathogens, from the dishes as you scrub them.

▶ If you see a greasy film on your dish water, it will be transferred to your dishes as you bring them out of the water. Rinsing will not be enough to remove the grease or contamination. If there is any organic material on the dishes, a sanitizer cannot be effective. The wash water must be changed frequently enough to prevent the dishes from being contaminated instead of cleaned.

Rinsing

▶ Some people prefer to rinse with running water and if you only have one sink, that may be the most efficient way to rinse your dishes.

▶ Whether you rinse your dishes in a sink full of hot water or under hot running water, the step of rinsing removes the soap residue and soil from the dishes. This step is important because without it, the dishes will not be clean and so cannot be effectively sanitized.

Sanitizing

▶ To sanitize your dishes, fill a sink with hot water, 110°F (44°C). Add one ounce (30 ml) of chlorine bleach per gallon (4 L) of water.

▶ You could use near-boiling hot water—180°F (82°C)—to sanitize your dishes instead of chlorine bleach or iodine. This temperature is far too high to be safely held in a hot water tank. You need to have a sink full of recently boiled water to achieve it and you need to protect yourself from scalding. There are dish racks with long handles designed for this purpose.

Hot Water Sanitizer 180°F (82°C)

Machine Wash Water 140°F (60°C)

Hand Wash Water 110°F (44°C)

▶ Place the clean, rinsed, dishes in the sanitizing solution or hot water. Make sure that there are no trapped air pockets. The sanitizer must make direct contact with the surface to be effective. Leave the dishes in the sanitizing solution for a few minutes.

▶ Do not rinse the dishes after they have been held in the sanitizing solution. Used in the correct ratios, both chlorine bleach and iodine will continue to work as the dishes dry and, as this happens, these solutions will evaporate from the surface of the dishes.

Air Drying

▶ Place the dishes in a dish rack in such a way that the water will drain off and allow them to air dry. If water is allowed to collect in dishes it could provide an environment for any pathogens (now introduced from the air or dust) to multiply on the surface of the dishes.

▶ Silverware or cutlery should be allowed to dry in a cutlery container with the blades of the knives, the tines of forks and bowls of the spoons exposed to the air. Avoid placing silverware in such a way that they nestle, as in the bowl of one spoon resting directly inside another spoon.

▶ This way of placing the cutlery, will allow the air to reach and thoroughly dry the parts that touch the mouth. It is important to prevent over-crowding so that there is room to reach your clean fingers down into the container and remove the cutlery by the handles, when you want to put them away.

▶ Another way of avoiding hand contact with the silverware parts that touch the mouths to pick up the container and empty the contents onto a sanitized surface or towel so that it can be easily picked up by the handles and stored later.

▶ It is not recommended to towel dry dishes. Pathogens from hands, or the towel could be transferred to the clean dishes.

► If you must towel dry dishes use only clean towels that are fresh from the laundry.

Washing Dishes By Machine

► Some detergents for home dish washing machines contain some sanitizer. The spring loaded cup of detergent opens in the last washing cycle, and can sanitize if the dishes were cleaned in the first washing cycle.

► It is important to pre-rinse dishes that will be cleaned in the machine. If there is any organic material left on the dishes, the sanitizer cannot be effective. The sanitizer will be used up on organic material and won't have a chance to work on any remaining pathogens.

► Some modern dish washing machines have soil separator traps which collect food particles to be flushed with each draining to reduce the amount of contaminants in the wash and rinse water.

► If there are particles of food left on dishes at the end of the process of wash, rinse and dry, the dishes are not "okay because they've been cleaned." Remember that pathogens can survive temperatures of up to 170°F (77°C) and the water will not be that hot. Detergent helps to lift off particles, but does not kill pathogens. Some bacteria form spores and they can be present in the dried food particles, waiting for growth conditions to improve.

Pre-rinse

Scrape away any chunks of food from the dishes and rinse away as much organic material as possible. The cleaner the wash water is, the cleaner the dishes will be.

Loading the Dishwasher

► Place the dishes in the racks in such a way that the water can reach all food contact surfaces and will drain off and allow them to air dry

► Avoid setting them in such a way that water and food particles can collect in them. That water could provide an environment for any pathogens (especially if there are any food particles) to multiply on the surface of the dishes.

▶ Silverware or cutlery should be placed in the cutlery container with the blades of the knives, the tines of forks and bowls of the spoons exposed to the water and detergent. Avoid placing the silverware in such a way it will nestle, as in the bowl of one spoon resting directly inside another spoon.

▶ This way of placing the cutlery, will allow the water to reach and thoroughly clean the parts that touch the mouth. It is important to prevent over-crowding so that there is room to reach your clean fingers down into the container and remove the cutlery by the handles, for storage later.

▶ Another way of avoiding hand contact with the clean cutlery is to pick up the container and empty the contents onto a sanitized surface or towel so that it can be easily picked up by the handles, and then stored.

Washing and Rinsing

Mechanical dishwashers do the job of washing and rinsing for us. The spray of water must reach all surfaces to clean effectively.

▶ Many detergents for machine dishwashers recommend high temperatures for wash and rinse water. The high temperature helps dissolve the detergent and prevent a soapy residue on your dishes. However, if you have small children or elderly people in your home, it is not recommended to set your hot water tank above 110°F (44°C). Hotter temperatures could cause accidental burns. Some dishwashers have a built-in booster to raise the water temperature to 140°F (60°C).

Hot Water Sanitizer
180°F (82°C)

Machine Wash Water
140° F (60°C)

Hand Wash Water
110°F (44°C)

▶ The wash water must be 140°F (60°C) or lower to prevent "baking on" food particles, while the rinse water can be as hot as you like.

▶ If the hot water in your tank has been used recently, there may not be enough hot water to do the job. It is best to wait for the tank to come back to the correct temperature to ensure that your dishes will be cleaned properly.

Sanitizing

There are many different models of dishwashers designed for domestic use. They are not always designed to sanitize the dishes.

► To sanitize dishes in the dishwasher, you need either a high heat of 180°F (82°C) or a chemical sanitizer after the rinse cycle.

► In many machines the air temperature in the drying cycle does not get the dishes hot enough to sanitize them, but removing moisture removes one of the important things the bacteria need to multiply.

► If the dishwasher is in good condition, loaded properly and hot water and detergent are used, a mechanical dishwasher generally will still get dishes cleaner than they would be if they were washed by hand.

Air Drying

Home dishwashers heat the air inside the machine during the drying cycle. This hot air in turn, heats the dishes and speeds the drying process.

If you want to save electricity, you may decide to set the machine so it turns off when it reaches the drying cycle. To allow the dishes to air dry, you would have the best results by leaving the door open for several hours, possibly overnight. The dishes should be completely dry before you store them.

Storing Clean Dishware

► All clean dishes, glassware and cutlery need to be stored in a way that they are protected from general contamination such as dust, sneezes, insects or mice.

► In kitchens that do not have doors on the dish storage cupboards, it is important to store the glasses and cups upside down to prevent contamination of their food contact areas from dust, insects or sneezes.

► It is not recommended to use absorbent materials such as cloth towels, or shelf paper to line storage cupboards. If there is any moisture on the glassware, it could be held in the towel or paper and any pathogens present could multiply on the dishes. If you need to store these items upside down, you could use a mesh-like rubber matting which allows for air circulation to continue the drying process.

▶ Any dishes stored in the open should be protected from dust, sneezes or contamination from food spills or kitchen splatter.

▶ All dishware should be stored in such a way that you do not have to touch the food contact surfaces to put them away or to take them out for use. You should be able to hold the glasses by the bottom rather than the lips; the plates by the edges; pots and casserole dishes by the handles; and cutlery and tools, like whisks, by the handles.

▶ If you store cutlery standing, in containers on the counter, the handles should be up to prevent hand contact with the parts that touch the mouth or food.

Single Service Items

Dishware like disposable plastic, styrofoam or paper plates, cups, glasses and cutlery are designed to be used only once. They cannot be cleaned effectively for re-use. The surfaces are easily scratched and these tiny scratches can hold pathogens. These materials cannot withstand the heat or chemicals that would sanitize them.

Re-using Plastic Containers/Bags

As we become more concerned with environmental issues, many people are making an effort to recycle or re-use plastic containers in an effort to reduce the amount of garbage sent to landfill sites. This approach can be appropriate if certain precautions are taken.

▶ If the container or bag has ever held a high protein food such as raw or cooked meat, it cannot be cleaned effectively enough to be safely re-used.

▶ If the container or bag has been allowed to hold foods that have spoiled or developed mold, it cannot be cleaned effectively enough to be safely re-used.

▶ If a container once held food, it should never be re-used to hold a chemical or cleaner. This can lead to serious problems if the chemical resembles a food substance, such as detergent which can be mistaken for flour, salt or powdered milk.

Pest Control

Throughout history, humans have had to combat diseases that have been spread as a result of insects and rodents. These pests can harm food supplies by eating them or by contaminating them with pathogens in their urine, feces or the surface of their bodies as they travel.

Pathogens can be picked up by insects and rodents when they crawl over or ingest decaying foods, garbage, sewage, or feces from humans or animals. When they nibble or walk on our food or counter top or leave fecal droppings, pathogens are transmitted to the food, shelf or utensils.

Cockroaches

Cockroaches are one of the most common insect pests. They are known to be carriers of pathogens such as *Salmonella*. These insects prefer dark, warm, moist areas. They may hide in cracks and crevices, behind or under refrigerators, stoves, dishwashers, in water drains and in corrugated cardboard. Cockroaches are fond of starchy and sweet materials, meat, bakery products, dead animals, plants, glue, hair and grease.

Signs of a serious cockroach infestation include a strong "musty" odor, feces which look like large grains of ground pepper, or oval egg casings which may be brown, red or black. Since cockroaches prefer to look for food and water in the dark, the presence of a cockroach in daylight is a sign of serious infestation.

Infestations are very difficult to control; dealing with them usually requires the assistance of someone who is trained in pest management.

Flies

Flies are found all over the world. Warm, moist, decaying organic matter is necessary for fly eggs to hatch and the larvae, or maggots, to grow, so it is important to keep these materials far away from your kitchen. Flies feed on human and animal waste and garbage. Pathogens found in these wastes can be transferred to their hairy bodies, feet, mouths and intestinal tracts.

Many species have "straw-like" mouth parts and so they need to liquefy food in order to ingest it. These flies "vomit" enzymes onto the food to dissolve it before eating. So if a fly lands on your lunch, counter top, or utensils it can deposit pathogens from its body, droppings or vomitus.

Rodents

Rats and mice are the rodents that cause the greatest concern for food safety in many parts of the world. They may carry a great number of pathogens because of their contact with decaying matter, garbage and sewage. They contaminate foods through pathogens transmitted from their hairy bodies, urine and feces.

Signs of a rodent infestation include:

► Droppings which look like elongated pellets.

► Gnaw marks, a rodent's teeth are continually growing and they need to gnaw to keep them short (they can easily chew through plastic containers, paperboard, wood and even porcelain and unhardened concrete).

► Rat burrows which are most often found in dirt near the foundation of a building.

► Damaged food that has been partially eaten,

► Nesting material such as scraps of paper, sawdust, fiberglass insulation or hair.

► Runways or tracks that may be visible on dusty surfaces along walls; rats leave a dark oily residue.

► Sounds of running or gnawing especially in vertical walls or false ceilings.

Control Measures

► The most important thing you can do to control the most serious pests is remove their food supply. Without access to food and water they cannot find what they need to survive in your kitchen or storage area.

► Keep your kitchen and storage areas clean. Remove any crumbs, spills or debris. Clean behind and under stoves, refrigerators, dishwashers, and storage shelves frequently.

▶ Keep your kitchen and storage areas dry. Pests can be attracted by water or moisture.

▶ Store foods in sealed containers. If you need to protect foods from rodents, glass or metal containers will be most effective.

▶ Screens on windows will help keep flies and rats from getting inside your kitchen or storage area. Some rats have been known to jump six to eight feet (1.8 to 2.4 meters).

▶ Seal holes around pipes and cracks and crevices in walls and foundations with materials such as metal or cement that will stand up to gnawing rats and mice.

▶ Placing screens over ventilation pipes leading to the outside and over floor drains can help to keep pests outside.

Pesticides

The use of pesticides always requires great care. Pesticides used near food, food storage areas, counter tops or sinks can be especially dangerous. Where there are occasional signs of insects and rodents, well located fly attractors, glue boards and rodent traps may help eliminate them.

▶ Fly paper is not recommended for the kitchen area since glue and flies may drip onto food and counter tops.

▶ Pesticides in a dust or granular form may be tracked by rodents or cockroaches into foods or onto dishware and counter tops where it could be mistaken for flour or sugar.

▶ Serious infestations should be handled by professional exterminators because they know the most effective and safest ways to control pests.

Sources of Foodborne Illness

Microbial Food Poisoning

Illness

Amoebic dysentry caused by the parasitic protozoa *Entamoeba histolytica.*

Food Source

Raw vegetables and fruits contaminated from untreated sewage infected by humans and animals.

Signs and Symptoms

Abdominal pain, constipation or diarrhea containing blood and mucous.

Contributing Factor

Inadequate cooking; infected people touching food; poor personal hygiene.

Prevention

Learn and practice good personal hygiene. Avoid handling food when ill—especially ready-to-eat foods. Wash fruits and vegetables.

Illness

Angiostrongyliasis caused by the parasitic roundworm *Parastrongylus cantonensis* (lung worm of rats).

Food Source

Raw crabs, prawns, fish, slugs, shrimp and snails that become infected by ingesting the larvae.

Signs and Symptoms

Diarrhea, abdominal pain, nausea, headache, stiff neck and back, and low grade fever.

Contributing Factor

Inadequate cooking of foods.

Prevention

Cook shellfish including mollusks and crustaceans thoroughly.

Illness

Anisakiasis caused by the parasitic roundworm *Anisakis*.

Food Source

Fish.

Associated Foods

Cod, haddock, fluke, Pacific salmon, herring and flounder; raw fish items such as sushi, ceviché, cold-smoked fish, and lightly cooked fillets.

Signs and Symptoms

Nausea, abdominal pain, may mimic appendicitis, coughing (larvae can migrate to the throat), and fever appear from one hour to two weeks after ingestion.

Contributing Factor

Eating inadequately cooked or raw fish.

Prevention

Obtain fish from a reputable source. Fish should be cleaned soon after they are caught to prevent migration of the worm from the stomach of the fish to the flesh. Fresh fish should be cooked to at least 145°F (63°C). Fish should be frozen at 0°F (-18°C) for seven days to destroy the worm and eggs if the fish will be used raw, as in sushi, or if only lightly cooked.

Illness

Bacillus cereus gastroenteritis caused by spore-forming *Bacillus cereus* bacteria. These organisms produce two different toxins as they multiply, each with differing onset times and symptoms.

Food Source

Spores are found in soil and form in foods such as rice when they are dried.

Associated Foods

Cereal products, rice, flour, starch, spices, dry-mix products, custards and sauces, meatloaf, vegetable sprouts, meats and milk.

Signs and Symptoms

Some reports of vomiting within 30 minutes to five hours after ingestion, resulting from one toxin. Nausea, abdominal pain, diarrhea within eight to 16 hours after ingestion, resulting from the other toxin. Recovery generally takes from one to several days.

Contributing Factor

Spores survive safe cooking temperatures and are activated when cooked or reconstituted foods are allowed to remain in the danger zone temperature range—i.e. preparing foods several hours before serving and not keeping the foods hot or refrigerating them or inadequately reheating leftovers. One of the toxins produced during bacterial multiplication is not destroyed by cooking temperatures.

Prevention

After cooking, keep potentially hazardous foods at either refrigeration or hot holding temperatures. Dense foods or large batches of foods must be cooled or reheated rapidly. If these foods have been held in the danger zone temperature range for more than two hours they should be thrown out.

Illness

Botulism is a severe intoxication caused by spore-forming *Clostridium botulinum* bacteria. The bacteria require an absence of oxygen to grow and the toxin is produced as they multiply. Usually found in canned, bottled, or vacuum packaged foods. Outbreaks are rare, but the mortality rate is high.

Food Source

Low-acid foods that come in contact with the soil—i.e vegetables or herbs, honey, bottom sediment of water bodies, intestinal tract of fish and mammals, crabs and other shellfish.

Associated Foods

Improperly canned or bottled low-acid foods such as green beans, asparagus, peppers, mushrooms etc.; vacuumed packed, unfrozen or uncooked smoked fish; onion or garlic products packed in oil; turkeyloaf; thick stew; and foil-wrapped baked potatoes.

Signs and Symptoms

Toxin acts on nerve cells and blocks their messages to the muscles—symptoms may begin as vomiting, diarrhea, or constipation, headache, dizziness,—progressing to double vision, weakness, difficulty swallowing and speaking, progressive respiratory paralysis, respiratory failure, and death, if not treated quickly. The first symptoms appear 12 to 36 hours or longer after eating the contaminated food.

Contributing Factor

Spores are resistant to high heat and a pressure canner is needed to reach hotter than boiling temperatures to destroy them. The toxins associated with fish can be produced at temperatures as low as 37°F (3°C). These bacteria will grow and produce toxins on many types of foods, including raw vegetables, if they are kept in an oxygen free environment.

Prevention

Buy canned or bottled foods in good condition. Don't use cans that are swollen or have pressure inside and don't use the food if the contents are foamy, smell foul or show any sign of spoilage. Buy vacuum packaged seafood only in a frozen state and thaw in the refrigerator with the package open, allowing the food to be exposed to air.

Illness

Brucellosis caused by *Brucella* bacteria.

Associated Foods

Raw milk and goat cheese.

Signs and Symptoms

Fever, chills, sweats, weakness, malaise, headache, muscle and joint pain, loss of weight.

Contributing Factor

Livestock infected with Brucella bacteria; failure to pasteurize milk.

Prevention

Buy foods produced from inspected herds. Use only pasteurized milk or products made from pasteurized dairy sources.

Illness

Campylobacteriosis caused by *Campylobacter jejuni* bacteria.

Food Source

Intestines of birds and animals.

Associated Foods

Raw meats, poultry, unpasteurized milk, contaminated shellfish, foods contaminated from untreated sewage infected by humans and animals.

Signs and Symptoms
　　Abdominal cramps, diarrhea.
Contributing Factor
　　Inadequate refrigeration; inadequate cooking and reheating; cross-contamination through inadequate cleaning of equipment; infected food handlers touching cooked food.
Prevention
　　Learn and practice good personal hygiene. Prevent cross-contamination—sanitize tools, equipment and work area after preparing raw meat. Protect ready-to-eat foods from contact with raw foods.

Illness
　　Cholera caused by *Vibrio cholerae* bacteria.
Food Source
　　Raw fish and shellfish, food washed or prepared with water contaminated from untreated sewage infected by humans and animals, untreated drinking water.
Signs and Symptoms
　　Profuse, watery diarrhea (rice-water stools), vomiting, abdominal pain, dehydration, thirst, collapse, reduced skin turgor and sunken eyes. The early symptoms appear within two to three days.
Contributing Factor
　　Obtaining fish and shellfish from sewage contaminated waters in endemic areas; poor personal hygiene of infected food handlers touching food; inadequate cooking; using contaminated water to wash or freshen food; inadequate sewage disposal; using human waste for fertilizer.
Prevention
　　Learn and practice good personal hygiene. Avoid handling food when ill, especially ready-to-eat foods. Use safe drinking water when washing or preparing food. Obtain seafood from a reputable supplier and cook shellfish thoroughly.

Illness
　　Ciguatera poisoning caused by a naturally occurring toxin produced by several species of algae. Fish eat other marine life that contain the toxin and become toxic themselves.

Food Source

Numerous varieties of tropical fish.

Signs and Symptoms

Tingling and numbness about the mouth, metallic taste, dry mouth, gastrointestinal symptoms, watery stools, muscular pain, dizziness, dilated pupils, blurred vision, prostration, paralysis.

Contributing Factor

Eating liver, intestines, roe, gonads, or flesh of topical reef fish; usually large reef fish are more commonly toxic.

Prevention

Obtain marine fish from a reputable source. Avoid consuming liver, intestines, roe and gonads of tropical marine fish.

Illness

Clostridium perfringens gastroenteritis caused by spore-forming *Clostridium perfringens* bacteria. This is one of the most common foodborne illnesses reported in North America. Illness is caused by toxin production in the digestive tract.

Food Source

Found throughout the environment, particularly in soil, gastrointestinal tracts of humans and animals, and likely to be found on any raw food.

Associated Foods

Cooked meat, poultry, gravy sauces, chili, stews, and soups.

Signs and Symptoms

Intense abdominal pain and explosive diarrhea occur eight to 24 hours after ingestion. The illness is usually over within 24 hours.

Contributing Factor

Spores survive safe cooking temperatures and are activated when cooked or reconstituted foods are allowed to remain in the danger zone temperature range—i.e. preparing foods several hours before serving and not keeping the foods hot or refrigerating them or inadequate reheating of leftovers.

Prevention

After cooking, keep potentially hazardous foods at refrigeration or hot holding temperatures. If these foods have been held in the danger zone temperature range for more than two hours they should be thrown out.

Illness

Cryptosporidiosis caused by the parasitic protozoa *Cryptosporidium parvum*. Infection can cause serious illness in immuno-compromised persons.

Food Source

Foods contaminated from untreated sewage infected by humans and animals; untreated drinking water.

Signs and Symptoms

Profuse, watery diarrhea, abdominal pain; in children these symptoms may be preceded by a loss of appetite, weight loss and vomiting. Symptoms appear from two to 14 days after ingestion and generally last less than 30 days. Some infected people do not develop symptoms.

Contributing Factor

Infected people may continue to shed the pathogen for several weeks after the symptoms pass.

Prevention

Protect against contamination of public water supplies by filtration designed to remove cysts. Learn and practice good personal hygiene. Wash all fruits and vegetables before eating them.

Illness

Cyclospora infection caused by the parasitic protozoa *Cyclospora cayentanensis*.

Food Source

Foods contaminated from untreated sewage infected by humans and animals; raw fruits; untreated drinking water.

Signs and Symptoms

Watery diarrhea, loss of appetite, weight loss, bloating, increased gas, abdominal pain, nausea, vomiting, tiredness, muscle aches, and low grade fever and can last from nine to 43 days. Some infected people do not develop symptoms.

Contributing Factor

Infected people may continue to shed the pathogen for several weeks after the symptoms pass.

Prevention

Protect against contamination of public water supplies by filtration designed to remove cysts. Learn and practice good personal hygiene. Wash all fruits and vegetables before eating them.

Illness

Escherichia coli gastroenteritis caused by *Escherichia coli 0157:H7* bacteria. Also referred to as *E. coli*, it has recently emerged as an important pathogen, and acquired notoriety because of the severity of the illness that has been called "hamburger disease," though its sources are not limited to hamburgers.

Food Source

Undercooked or raw ground meat, especially beef, but also pork and lamb, various other foods and water contaminated from untreated sewage infected by humans and animals.

Signs and Symptoms

Abdominal pain, nausea, headache, muscular pain, severe cramping, diarrhea, initially watery, may become bloody, occasional vomiting. Fever is low-grade or absent. Children are particularly vulnerable and can suffer kidney failure—potentially fatal. Symptoms appear between 12 and 60 hours following ingestion and last for an average of eight days.

Contributing Factor

Infected food handlers touching food; inadequate cooking; cross-contamination from inadequate cleaning and disinfection of equipment.

Prevention

Learn and practice good personal hygiene. Prevent cross-contamination—sanitize tools, equipment and work area after preparing raw meat. Protect ready-to-eat foods from contact with raw foods. Cook hamburgers to a well-done stage—if small children will eat them, cut them open to make certain they cooked thoroughly (juices should run clear and the meat should not be pink). Keep potentially hazardous foods at refrigeration or hot holding temperatures. If these foods have been held in the danger zone temperature range for more than two hours they should be thrown out.

Illness

Giardiasis (Beaver Fever) caused by the parasitic protozoa *Giardia lamblia*.

Food Source

Foods contaminated from untreated sewage infected by humans and animals, raw vegetables and fruits, untreated drinking water from streams and rivers.

Signs and Symptoms

Abdominal pain, mucoid diarrhea, fatty stools, bloating, fatigue and weight loss. Early symptoms appear, on average, 10 days after ingestion.

Contributing Factor

Poor personal hygiene of infected food handlers touching food; inadequate cooking; inadequate sewage disposal, contaminated water.

Prevention

Protect against contamination of public water supplies by filtration designed to remove cysts. Learn and practice good personal hygiene. Purchase ice from approved sources.

Illness

Infectious Hepatitis caused by the *Hepatitis A* virus.

Food Source

Shellfish; any food contaminated from untreated sewage infected by humans and animals; untreated drinking water.

Signs and Symptoms

Fever, malaise, lassitude, anorexia, nausea, abdominal pain, jaundice. The incubation period is from three to six weeks and the illness lasts from a few weeks to a few months.

Contributing Factor

The pathogen can be shed by the infected person during the long incubation period and one week after jaundice appears. Infected food handlers touching food; poor personal hygiene; inadequate cooking; harvesting shellfish from sewage contaminated waters; inadequate sewage disposal.

Prevention

Learn and practice good personal hygiene. Obtain seafood from reputable sources. Cook shellfish thoroughly.

Illness

Listeriosis caused by *Listeria monocytogenes* bacteria. Infection can have severe medical consequences, especially for pregnant women who can transmit the disease to the fetus. Immuno-compromised people can become ill with meningitis.

Food Source

Infected humans, wild and domestic mammals, poultry, soil, and untreated drinking water.

Associated Foods

Raw milk and ice cream and soft-ripened cheese (especially made with unpasteurized milk), raw vegetables, raw-meat sausages, raw and cooked poultry, raw meats (all types), raw and smoked fish.

Signs and Symptoms

Nausea, vomiting, headache appear from a few days to 3 weeks after ingestion.

Contributing Factor

These bacteria are noteworthy because they continue to grow at temperatures as low as 32.5°F (0.3°C) and can grow on wet floors, in drains, in ceiling condensation and sponges and cloths.

Prevention

Use pasteurized dairy products. Use foods by the "best before" date. Prevent cross-contamination—sanitize tools, equipment and work areas and keep them dry.

Illness

Norwalk **virus** is believed to be transmitted primarily by humans and can survive in food for an extended period of time.

Food Source

Foods or water contaminated from untreated sewage infected by humans and animals.

Associated Foods

Raw shellfish—especially clams and oysters, cole slaw, raw vegetables, commercially manufactured ice cubes.

Signs and Symptoms

Nausea and vomiting, diarrhea, abdominal pain, headache, low-grade fever occur from 24 to 48 hours after ingestion and can last from 24 to 48 hours.

Contributing Factor

Obtaining shellfish from sewage contaminated waters in endemic areas; poor personal hygiene of infected food handlers touching food; inadequate cooking.

Prevention

Learn and practice good personal hygiene. Purchase seafood from approved sources. Cook shellfish thoroughly. Obtain seafood from a reputable source. Use potable water for drinking and ice. Avoid cross-contamination.

Illness

Paralytic Shellfish Poisoning caused by naturally occurring Saxitoxins produced by blooming algae (red tide).

Food Source

Mussels and clams.

Signs and Symptoms

Tingling, burning, numbness around lips and finger tips, giddiness, incoherent speech, respiratory paralysis.

Contributing Factor

Harvesting shellfish from waters with high concentrations of Saxitoxin.

Prevention

Obtain shellfish from a reputable source. Observe warnings for closed area if harvesting shellfish personally.

Illness

Salmonellosis caused by *Salmonella* bacteria. These bacteria have been the source of foodborne illness worldwide. Over 2,000 serotypes have been identified, producing infections classified as mild to severe.

Food Source

Poultry and meat and their products, eggs and egg products, other foods contaminated by various serotypes of salmonellae from water contaminated with untreated sewage infected by humans and animals.

Associated Foods

Most commonly poultry, eggs, meats, and meat products; but has been associated with salad greens, alfalfa and bean sprouts, sliced tomatoes, cantaloupe, melons, milk and drinking water (Typhoid Fever), sewage.

Signs and Symptoms

Abdominal pain, diarrhea, chills, fever, nausea, vomiting, malaise.

Contributing Factor

Infected people may continue to shed the pathogen for several weeks after the symptoms pass. The bacteria survive for long periods in frozen and in dehydrated foods; inadequate refrigeration; inadequate cooking and reheating; cross-contamination through inadequate cleaning of equipment; infected food handlers touching cooked food; improper waste disposal; obtaining food from contaminated sources; bacteria can form in the ova of poultry and are present inside the egg before it is laid. (*Salmonella enteritidis*).

Prevention

Learn and practice good personal hygiene. Prevent cross-contamination—sanitize tools, equipment and work area after preparing raw meat. Protect ready-to-eat foods from contact with raw foods. Use pasteurized egg products or cook eggs thoroughly for people at high risk from food poisoning. Keep potentially hazardous foods at refrigeration or hot holding temperatures. If these foods have been held in the danger zone temperature range for more than two hours they should be thrown out.

Illness

Scombroid poisoning caused by consuming histamines produced in scombroid fish after they are caught and in the production of certain cheese.

Food Source

Tuna, mackerel, Pacific dolphin, mahi-mahi, marlin and bluefish, cheese.

Signs and Symptoms

Headache, dizziness, nausea, vomiting, peppery taste, burning throat, facial swelling and flushing, stomach pain, itching skin.

Contributing Factor

Inadequate refrigeration of scombroid fish. Certain strains of bacteria which produce histamines allowed to contaminate cheese, due to unsanitary conditions during production.

Prevention

Obtain fish and other foods from a reputable source. Place scombroid fish on ice or freeze them immediately.

Illness

Shigellosis caused by *Shigella* bacteria.

Food Source

Any food washed or prepared with water contaminated from untreated sewage infected by humans and animals.

Associated Foods

Moist prepared foods, potato and other salads, raw fruits and vegetables, unpasteurized milk and dairy foods, poultry, and untreated drinking water.

Signs and Symptoms

Abdominal pain, diarrhea, bloody and mucoid stools, fever, diarrhea, cramps, chills, fever can appear between one and seven days after ingestion.

Contributing Factor

Infected people can shed the pathogen for several weeks; infected food handlers touching food; inadequate refrigeration; inadequate cooking and reheating.

Prevention

Learn and practice good personal hygiene. Keep potentially hazardous foods at refrigeration or hot holding temperatures. If these foods have been held in the danger zone temperature range for more than two hours they should be thrown out.

Illness

Staphylococcal intoxication caused by *Staphylococcus aureus* bacteria. Food becomes contaminated by the toxin produced by these bacteria as they multiply in the food. Because the toxin is heat stable, the bacteria must be prevented from growing in the first place.

Food Source

Any food contaminated by infected humans.

Associated Foods

Meat and poultry products, cured meats, fish, cream sauces, cream-filled pastry, food mixtures, leftover food.

Signs and Symptoms

Nausea, vomiting, retching, abdominal pain, diarrhea, prostration. Symptoms appear from one to eight hours after consuming the toxin. Recovery is generally within 24 hours.

Contributing Factor

These bacteria are found normally on the skin and in the nose of people everywhere. The toxin produced by these bacteria is heat stable and will not be destroyed by boiling or even pressure canning temperatures. Inadequate refrigeration; inadequate cooking and reheating; food handlers with skin infection or infected cuts touching cooked food; preparing food several hours before serving.

Prevention

Learn and practice good personal hygiene. Keep potentially hazardous foods at refrigeration or hot holding temperatures. If these foods have been held in the danger zone temperature range for more than two hours they should be thrown out.

Illness

Streptococcal infection caused by several groups of *Streptococcus* bacteria, two of which can be transmitted to humans in food. Group A has one species, *S. pyogenes*. Group D has five species, *S. faecalis, S. faecium, S. durans, S. avium,* and *S. bovis.*

Food Source

Any foods contaminated by infected humans.

Associated Foods

Raw milk, ice cream, custards, cheese, seafood salad, potato salad, food containing eggs and ham and meat sandwiches.

Signs and Symptoms

Group A symptoms include sore and red throat, pain on swallowing, tonsillitis, high fever, headache, nausea, vomiting, malaise, rhinorrhea, occasionally a rash. Group D symptoms include diarrhea, abdominal cramps, nausea and vomiting, fever, chills, dizziness.

Contributing Factor

Food handlers with skin infection or infected cuts touching cooked food; inadequate refrigeration; inadequate cooking or reheating; preparing food several hours before serving.

Prevention

Learn and practice good personal hygiene. Keep potentially hazardous foods at refrigeration or hot holding temperatures. If these foods have been held in the danger zone temperature range for more than two hours they should be thrown out.

Illness

Tapeworm infection—beef

Food Source

Raw or insufficiently cooked beef.

Signs and Symptoms

Vague discomfort, hunger pain, loss of weight, abdominal pain.

Contributing Factor

Lack of meat inspection; inadequate cooking; inadequate sewage disposal; sewage contaminated pastures.

Prevention

Obtain beef from a reputable source. As a precaution beef can be frozen at 0°F (-18°C) for seven days to destroy the worm and eggs.

Illness

Tapeworm infection—fish

Food Source

Raw or insufficiently cooked freshwater fish.

Signs and Symptoms

Vague gastrointestinal discomfort, anemia may occur.

Contributing Factor

Inadequate cooking; inadequate sewage disposal; sewage contaminated lakes.

Prevention

Obtain fish from a reputable source. As a precaution fish can be frozen at 0°F (-18°C) for seven days to destroy the worm and eggs.

Illness

Tapeworm infection—pork

Food Source

Raw or insufficiently cooked pork.

Signs and Symptoms

Vague discomfort, hunger pains, loss of weight.

Contributing Factor

Lack of meat inspection; inadequate cooking, inadequate sewage disposal; sewage contaminated pastures.

Prevention

Obtain pork from a reputable source. As a precaution pork can be frozen at 0°F (-18°C) for seven days to destroy the worm and eggs.

Illness

Tetradon poisoning caused by consuming fish contaminated with Tetradotoxin.

Food Source

Puffer-type fish.

Signs and Symptoms

Tingling sensation of fingers and toes, dizziness, pallor, numbness of mouth and extremities, gastrointestinal symptoms, hemorrhage and flaking of skin, eyes fixed, twitching, paralysis, cyanosis.

Contributing Factor

Eating puffer-type fish; failure to effectively remove intestines and gonads from puffer-type fish if they are to be eaten.

Prevention

Eating puffer-fish is not recommended.

Illness

Toxoplasmosis caused by the parasitic protozoa *Toxoplasma gondi*. Pregnant women can pass the infection to the fetus causing health problems, birth defects, miscarriage or stillbirth. Infection can cause serious illness in immuno-compromised persons.

Food Source

Foods washed or prepared with water contaminated from untreated sewage infected by humans and animals, drinking water.

Associated Foods

Raw or insufficiently cooked meat (rare), raw milk, untreated drinking water.

Signs and Symptoms

Fever, headache, sore muscles, tiredness, and sometimes blurred vision or loss of vision.

Contributing Factor

Inadequate cooking of meat; using unpasteurized dairy products.

Prevention

Protect against contamination of public water supplies by filtration designed to remove cysts. Obtain meats from reputable source. Use pasteurized dairy products. Learn and practice good personal hygiene.

Illness

Trichinosis caused by the parasitic roundworm *Trichinella spiralis.*

Food Source

Pork, bear meat, walrus flesh.

Signs and Symptoms

Abdominal pain, diarrhea, vomiting, fever, edema about eyes, muscular pain, chills, prostration, labored breathing.

Contributing Factor

Eating raw or inadequately cooked pork or bear meat; inadequate cooking or heat processing; feeding uncooked or inadequately heat processed garbage to hogs.

Prevention

Obtain meats from a reputable source, freeze pork for 20 to 30 days at °F (-18°C) or cook to a well done stage.

Illness

Typhoid fever caused by *Salmonella typhi* bacteria.

Food Source

Foods washed or prepared with water contaminated with untreated sewage infected by humans and animals; untreated drinking water.

Associated Foods

Shellfish, food contaminated by infected humans, raw milk, cheese, watercress, water.

Signs and Symptoms

Malaise, headache, fever, cough, nausea, vomiting, constipation, abdominal pain, chills, rose spots on body trunk, bloody stools.

Contributing Factor

Infected food handlers touching food; poor personal hygiene; inadequate cooking; inadequate refrigeration; inadequate sewage disposal; obtaining food from unsafe sources; harvesting shellfish from sewage-contaminated waters.

Prevention

Learn and practice good personal hygiene. Keep potentially hazardous foods at refrigeration or hot holding temperatures. If these foods have been held in the danger zone temperature range for more than two hours they should be thrown out. Use potable drinking water to prepare foods. Obtain seafood from a reputable supplier and cook shellfish thoroughly.

Illness

Vibrio parahaemolyticus gastroenteritis caused by *Vibrio parahaemolyticus* bacteria.

Food Source

Raw seafood—especially shellfish.

Signs and Symptoms

Abdominal pain, diarrhea, nausea, vomiting, fever, chills and headache occur an average of 15 hours after ingestion and last for three days.

Contributing Factor

Inadequate cooking; inadequate refrigeration; cross-contamination resulting from inadequate cleaning of equipment; using sea water in food preparation.

Prevention

Obtain seafood from a reputable supplier and cook shellfish thoroughly.

Illness

Vibrio vulnificus gastroenteritis caused by *Vibrio vulnificus* bacteria. Produces septicemia in people with liver disease, chronic alcoholism or those who are immuno-compromised.

Food Source

Raw seafood—especially shellfish.

Signs and Symptoms

Abdominal pain, diarrhea, nausea, vomiting and fever occur 12 hours to three days after ingestion. Over 50% of individuals with primary septicema die. Wounds sustained in coastal water can become infected, causing serious cell damage or death.

Contributing Factor

Inadequate cooking or inadequate refrigeration.

Prevention

Obtain seafood from a reputable supplier and cook shellfish thoroughly.

Illness

Viral gastroenteritis caused by *Enteric* viruses.

Food Source

Foods washed or prepared with water contaminated from untreated sewage infected by humans and animals or harvested from contaminated waters.

Signs and Symptoms

Diarrhea, fever, vomiting, abdominal pain and sometimes respiratory symptoms.

Contributing Factor

Poor personal hygiene of infected food handlers touching food; inadequate cooking and reheating.

Prevention

Learn and practice good personal hygiene. Obtain foods from a reputable source.

Illness

Yersiniosis caused by *Yersinia enterocolitica* bacteria.

Food Source

Soil, water, and is spread by animals such as pigs, cats, dogs and rodents.

Associated Foods

Meats—especially pork but also beef, lamb, tofu, oysters, fish, unpasteurized milk, and ice cream.

Signs and Symptoms

Acute watery diarrhea, vomiting, abdominal pain, fever and can mimic appendicitis. Symptoms generally appear within one to three days and last from days to weeks.

Contributing Factor

Continues to grow at temperatures as low as 32.5°F (0.3°C).

Prevention

Learn and practice good personal hygiene. Prevent cross-contamination—sanitize tools, equipment and work area. Protect ready-to-eat foods from contact with raw foods.

Toxic Mushrooms

Illness

Cyclopeptide and gyromitrin groups of mushroom poisoning.

Food Source

Amanita phalloides, Amanita erna, Galerina antumnalis, Gyromitra esculenta (false morels) and similar species of mushroom.

Signs and Symptoms

Abdominal pain, feeling of fullness, vomiting, protracted diarrhea, loss of strength, thirst, muscle cramps, feeble rapid pulse, collapse, jaundice, drowsiness, dilated pupils, coma, death.

Contributing Factor

Eating unknown varieties of mushrooms or mistaking toxic mushrooms for edible varieties.

Prevention

Obtain mushrooms from a reputable source. Only qualified people should harvest wild mushrooms.

Illness

Gastrointestinal irritating group of mushroom poisoning.

Food Source

Wild mushrooms.

Signs and Symptoms

Nausea, vomiting, retching, diarrhea, abdominal cramps.

Contributing Factor

Eating unknown varieties of mushrooms or mistaking toxic mushrooms for edible varieties.

Prevention

Obtain mushrooms from a reputable source. Only qualified people should harvest wild mushrooms.

Illness

Ibotenic acid group of mushroom poisoning.

Food Source

Amanita muscaria, Amanita pantherina and related species of mushrooms.

Signs and Symptoms

Drowsiness and state of intoxication, confusion, muscular spasms, delirium, visual disturbances.

Contributing Factor

Eating unknown varieties of mushrooms or mistaking toxic mushrooms for edible varieties.

Prevention

Obtain mushrooms from a reputable source. Only qualified people should harvest wild mushrooms.

Illness

Muscarine group of mushroom poisoning.

Food Source

Clitocybe dealbata, Clitocybe rivulosa and many species of Inocybe and Boletus mushrooms.

Signs and Symptoms

Excessive salivation, perspiration, tearing, reduced blood pressure, irregular pulse, pupils constricted, blurred vision, asthmatic breathing.

Contributing Factor

Eating unknown varieties of mushrooms or mistaking toxic mushrooms for edible varieties.

Prevention

Obtain mushrooms from a reputable source. Only qualified people should harvest wild mushrooms.

Chemical Food Poisoning

Illness
Antimony poisoning
Food Source
High-acid food and beverages stored in gray enamelware.
Signs and Symptoms
Vomiting, abdominal pain, diarrhea.
Contributing Factor
Using antimony-containing utensils, storing high-acid foods in gray enamelware.
Prevention
Use utensils made from food quality materials such as stainless steel, food-grade plastic or glass.

Illness
Cadmium poisoning
Food Source
High-acid food and beverages, candy love beads or cake decorations.
Signs and Symptoms
Nausea, vomiting, abdominal cramps, diarrhea, shock.
Contributing Factor
Storing high-acid foods in cadmium-containing containers, ingesting cadmium-containing foods, placing high-acid foods directly on shelving in refrigerators manufactured before the mid-1960's.
Prevention
Use utensils made from food quality materials such as stainless steel, food-grade plastic or glass. Do not use shelves from old refrigerators as cooking grills.

Illness
Copper poisoning
Food Source
High-acid food and beverages exposed to copper, copper in pipes and utensils.
Signs and Symptoms
Metallic taste, nausea, vomiting (green vomitus), abdominal pain, diarrhea.

Contributing Factor
  Storing high-acid foods in copper utensils or using copper pipes for dispensing high-acid beverages, faulty backflow preventor valves in beverage vending machines.
Prevention
  Use utensils made from food quality materials such as stainless steel, food-grade plastic or glass. Use approved tubing for dispensing high-acid or carbonated beverages.

Illness
  Fluoride poisoning
Food Source
  Any food accidentally contaminated with fluoride, particularly dry food such as dry milk, flour, baking powder and cake mixes.
Signs and Symptoms
  Salty or soapy taste, numbness of mouth, vomiting, diarrhea, abdominal pain, pallor, cyanosis, dilated pupils, spasms, collapse, shock.
Contributing Factor
  Storing insecticides in same area as food; mistaking pesticides for powdered food.
Prevention
  Store pesticides separate from food. Pay very careful attention when using insecticides in the kitchen or food storage areas. Use a professionally trained exterminator to handle infestations.

Illness
  Lead poisoning
Food Source
  High-acid food and beverages stored in lead-containing vessels, drinking water contaminated by lead soldered pipes, foods served or prepared in ceramic dishes containing lead, or any accidentally contaminated food.
Signs and Symptoms
  Metallic taste, burning of mouth, abdominal pain, milky vomitus, bloody or black stools, foul breath, shock, blue gum line.

Contributing Factor

Storing high-acid food in lead-containing vessels; using lead soldered pipes for transporting water, using ceramic dishes containing lead, storing pesticides in same area as food.

Prevention

Use utensils made from food quality materials such as stainless steel, food-grade plastic or glass. Use approved materials for piping drinking water.

Illness

Mercury poisoning

Food Source

Grains treated with mercury-containing fungicide; pork, fish and shellfish exposed to mercury compounds.

Signs and Symptoms

Numbness, weakness of legs, spastic paralysis, impairment of vision, blindness, coma.

Contributing Factor

Streams polluted with mercury compounds, feeding animals grains treated with mercury fungicides, eating mercury treated grains or meat from animals fed with such grains.

Prevention

Obtain foods from a reputable source.

Illness

Nitrite poisoning

Food Source

Cured meats, and any accidentally contaminated food.

Signs and Symptoms

Nausea, vomiting, cyanosis, headache, dizziness, weakness, loss of consciousness, chocolate brown colored blood.

Contributing Factor

Using excessive amounts of nitrites or nitrates for curing food, mistaking nitrites for common salt and other condiments.

Prevention

Obtain food from a reputable source.

Illness

Sodium hydroxide poisoning

Food Source

Bottled beverages when bottles have been cleaned with caustic cleaners.

Signs and Symptoms

Burning of lips, mouth and throat; vomiting, abdominal pain, diarrhea.

Contributing Factor

Improperly rinsed bottle used to store beverages.

Prevention

Thorough rinsing of bottles which have been cleaned with caustic washing compounds or detergents. Purchase beverages from a reputable source.

Illness

Tin poisoning

Food Source

High-acid food and beverages stored in tin-lined cans.

Signs and Symptoms

Bloating, nausea, vomiting, abdominal cramps, diarrhea, headache.

Contributing Factor

Using uncoated tin containers for storing high-acid foods.

Prevention

Use utensils made from food quality materials such as stainless steel, food-grade plastic or glass.

Illness

Zinc poisoning

Food Source

High-acid food and beverages stored in zinc galvanized containers.

Signs and Symptoms

Pain in mouth and abdomen, nausea, vomiting, dizziness.

Contributing Factor

Storing high-acid food in galvanized cans.

Prevention

Use utensils made from food quality materials such as stainless steel, food-grade plastic or glass.

Food Storage Chart

For more information check the **"best before"** date on the package.

CUPBOARD
(room temperature)

Unless otherwise specified,
times apply to unopened
packages

CEREAL GRAINS
(once opened, store in airtight
containers, away from light
and heat)

Bread crumbs (dry)	3 mos
Cereal (ready to eat)	8 mos
Cornmeal	6-8 mos
Crackers	6 mos
Pasta	several yrs
Rice	several yrs
Rolled oats	6-10 mos
White flour	1 yr
Whole wheat flour	3 mos

CANNED FOODS
(once opened, store covered
in refrigerator)

Evaporated milk	9-12 mos
Other canned foods	1 yr

DRY FOODS
(once opened, store in airtight
containers, away from light
and heat)

Baking powder, baking soda	1 yr
Beans, peas and lentils	several yrs

Chocolate (baking)	7 mos
Cocoa	10-12 mos
Coffee (ground)	1 mo
Coffee (instant)	1 yr
Coffee whitener	6 mos
Fruit (dried)	1 yr
Gelatin	1 yr
Jelly powder	2 yrs
Mixes (cake, pancake, tea biscuit)	1 yr
Mixes (pudding and pie filling)	18 mos
Mixes (main dish accompaniments)	9-12 mos
Potatoes (flakes)	1 yr
Skim milk powder	
- unopened	1 yr
- opened	1 mo
Sugar (all types)	several yrs
Tea bags	1 yr

MISCELLANEOUS FOODS

Honey	18 mos
Jams, jellies (once opened, store covered in refrigerator)	1 yr
Mayonnaise, salad dressings	
- unopened	6 mos
- opened (store covered in refrigerator)	1-2 mos
Molasses	2 yrs
Nuts	2 mos

Peanut butter
- unopened   6 mos
- opened   2 mos
Pectin
- powdered   2 yrs
- liquid   1 yr
- opened (store
covered in
refrigerator)   1 mo
Syrups
- maple, corn, table   1 yr
Vegetable oils   1 yr
- Virgin olive oil
(once opened, store
covered in
refrigerator)   1 yr
Vinegar   several yrs
Yeast (dry)   1 yr

VEGETABLES
Potatoes, rutabaga,
squash   1 wk
Tomatoes   1 wk

COOL ROOM
45-50°F, (7-10°C)
Apples   1 wk
Potatoes (mature)   6 mos
Onions
(dry, yellow skin)   6 wks
Squash (winter)   several mos
Rutabaga
(waxed)   several mos

REFRIGERATOR
(40°F, 4C) unless otherwise
specified, cover all foods
DAIRY PRODUCTS
AND EGGS
(check "best before" dates)
Butter - unopened   8 wks
- opened   3 wks
Cheese - cottage
(once opened)   3 days
- firm   several mos
- processed
(unopened)   several mos
- processed
(opened)   3-4 wks
Margarine - unopened   8 mos
- opened   1 mo
Milk, cream, yogurt
(once opened)   3 days
Eggs   3 wks

FISH AND SHELLFISH
Fish (cleaned) - raw   2-3 days
- cooked   1-2 days
Crab, clams, lobster,
mussels (live)   12-24 hrs
Oysters (live, in
seawater)   several days
Scallops, shrimp
(raw)   1-2 days
Shellfish (cooked)   1-2 days

FRESH FRUIT (ripe)
Apples   2 mos
Apricots (store
uncovered)   1 wk
Blueberries
(store uncovered)   2 days

Cherries	3 days
Rhubarb	1 wk
Cranberries	
(store uncovered)	1 wk
Grapes	5 days
Peaches	
(store uncovered)	1 wk
Pears	
(store uncovered)	1 wk
Plums	5 days
Raspberries	
(store uncovered)	2 days
Strawberries	
(store uncovered)	2 days

FRESH VEGETABLES

Asparagus	5 days
Beans (green, wax)	5 days
Beets	3-4 wks
Broccoli	1 wk
Brussel sprouts	1 wk
Cabbage	2 wks
Celery	2 wks
Carrots	several wks
Cauliflower	10 days
Corn	10 days
Peas	1 wk
Cucumbers	1 wk
Lettuce	1 wk
Mushrooms	5 days
Sprouts	2 days
Onions (green)	1 wk
Parsnips	several wks
Peppers (green, red)	1 wk
Potatoes (new)	1 wk
Spinach	2 days
Squash (summer)	1 wk

MEAT, POULTRY
Uncooked

Cured or smoked	6 - 7 days
Ground meat	1 - 2 days
Poultry	2 - 3 days
Roasts	3 - 4 days
Steaks, chops	2 - 3 days
Variety meats, giblets	1 - 2 days

Cooked

All meats and poultry	3 - 4 days
Casseroles, meat pies, meat sauces	1 - 2 days
Soups	1 - 2 days

MISCELLANEOUS FOODS

Coffee (ground)	2 mos
Nuts	4 mos
Whole wheat flour	3 mos
Shortening	12 mos

FREEZER (0°C, -18°C)
Use freezer wrapping or airtight containers. Freeze fresh food at its peak condition.

DAIRY PRODUCTS AND FATS

Butter - salted	1 yr
- unsalted	3 mos
Cheese - firm, processed	3 mos
Cream - table, whipping	1 mos
Ice Cream	1 mos
Margarine	6 mos
Milk	6 wks

FISH AND SHELLFISH

Fish (fat species: salmon, mackerel, lake trout)	2 mos
(lean species: cod, haddock, pike, smelt)	6 mos
Shellfish	2 - 4 mos

FRUITS AND VEGETABLES 1 yr

MEAT, POULTRY AND EGGS
Uncooked

Beef (roasts and steaks)	10 - 12 mos
Chicken, turkey - cut up	6-9 mos
- whole	1 yr
Cured or smoked meat	1 - 2 mos
Duck, goose	3 mos
Eggs (whites, yolks)	4 mos
Ground meat	3 mos
Lamb (chops, roasts)	8 - 12 mos
Pork (chops, roasts)	6 mos
Sausages, wieners	2 - 3 mos
Variety meats, giblets	3 - 4 mos
Veal (chops, roasts)	6 mos

Cooked

All meat	2 - 3 mos
All poultry	1 - 3 mos
Casseroles, Meat pies	3 mos

MISCELLANEOUS FOODS

Bean, pea, lentil casseroles	3 - 6 mos
Breads (yeast, baked or unbaked)	1 mo
Cakes, cookies (baked)	3 mos
Herbs	1 yr
Pastries, quick bread (baked)	1 mo
Pastry crust (unbaked)	2 mos
Pie (fruit, unbaked)	6 mos
Sandwiches	6 wks
Soups (stocks, cream)	3 mos

Glossary of FOODSAFE Terms and Rules

ADDITIVE (FOOD)   a substance added to food during processing or preparation, for example, Monosodium Glutamate. Food additives may become a chemical hazard if improperly used.

AEROBIC   the presence of free oxygen (e.g. exposed to air).

APPROVED   acceptable to the Health Department.

ANAEROBIC   the absence of free oxygen (e.g. in a vacuum or partial vacuum).

ANISAKIS   a roundworm parasite found in raw fish that causes the foodborne illness called Anisakiasis.

BACILLUS CEREUS   a facultative, pathogenic microorganism bacteria that forms spores. It produces toxins that cause foodborne intoxications of the same name.

BACTERIA   microorganisms which reproduce by simple cell division.

BACTERICIDE   an agent that causes death to bacteria cells but may not affect their spores.

CAMPYLOBACTER JEJUNI   an anaerobic pathogenic bacteria that does not form spores. It causes the foodborne infection called campylobacteriosis.

CARCINOGENIC   cancer producing.

CARRIER
a person or animal that harbors a pathogenic microorganism without showing apparent symptoms of disease. The carrier can be a source of transmission of the organism to food or others.

CEVICHE
a raw, marinated fish dish in South America. (see *Anisakis*)

CHEMICAL HAZARD
a risk to food safety by contamination with chemical substances, e.g. cleaning compounds, toxic metals, pesticides, etc.

CIGUATERA
a foodborne illness, usually involves a combination of gastrointestinal, neurological and cardiovascular disorders. Caused by the consumption of tropical marine finfish which have accumulated naturally occurring toxins through their diet. The toxins are known to originate from several species of algae.

CLEAN or CLEANSE
the absence of visible contaminants including dirt, food particles, grease, soil and other foreign material.

CLOSTRIDIUM BOTULINUM
an anaerobic pathogenic bacteria found in soil and in marine environments that forms spores and causes the often fatal foodborne intoxication called botulism.

CLOSTRIDIUM PERFRINGENS
an anaerobic pathogenic bacteria found in soil and dust, that forms spores and causes the foodborne toxicoinfection of the same name.

CONTAMINATION — the presence of harmful chemicals, foreign materials or pathogenic microorganisms or their toxins in food or drink.

CROSS-CONTAMINATION — occurs when pathogens are transfererred from a raw food or an infected person to a food, or to a surface such as a counter top, cutting board, utensil or dish and then to food.

CRYPTOSPORIDIUM — a parasite present in many herd animals (cows, goats and sheep). Some strains appear to be adapted to certain humans.

CYCLOSPORA CAYETANENSIS — a parasite composed of one cell. It may be transmitted through infected feces by the fecal-oral route by drinking contaminated water or eating contaminated food.

DANGER ZONE — the temperature range between 40°F to 140°F (4°C to 60°C). Pathogens and spoilage organisms grow rapidly in this zone.

DETERGENTS — a group of chemical products used for cleaning food contact surfaces, dishware, equipment, etc.

DRY STORAGE — an area or room used for storing non-perishable foods.

ESCHERICHIA (E) COLI 0157 AND OTHER TYPES

a facultative pathogenic bacteria that is non-spore forming. In the case of the 0157:H7 species it can cause the foodborne toxicoinfection Hemolytic Uremic Syndrome also known as barbecue syndrome or hamburger disease.

ENZYME

a protein able to initiate or accelerate specific chemical reactions in the metabolism of plants and animals; an organic catalyst.

FACULTATIVE

microorganisms that can grow either with or without the presence of free oxygen.

F.I.F.O. RULE

the principle of food stock and storage rotation, e.g. First In, First Out.

FOOD

any edible substance whether raw, cooked or processed, including water and ice.

FOODBORNE INFECTION

illness caused by eating contaminated food.

FOODBORNE INTOXICATION

illness caused by eating food containing toxins produced by pathogenic microorganisms.

FOODBORNE TOXICO-INFECTION

illness caused by eating food contaminated with pathogenic microorganisms that produce infection in the host, they also form toxins that are produced after ingestion of the pathogen which in turn cause intoxication, e.g. *E. Coli 0157:H7* or *C. perfringens*.

FOOD CONTACT SURFACE	any part of utensils or equipment with which food normally comes into contact during transportation, storage, preparation or service.
FUNGI	yeasts, molds and mushrooms, of which some are pathogenic to man.
GASTROENTERITIS	inflammation of the linings of the digestive tract which may be a result of eating contaminated food.
GIARDIA LAMBLIA	a pathogenic microorganism (protozoa) that is parasitic in humans and other warm-blooded animals. It causes the foodborne infection Giardiasis (also known as Beaver Fever).
HACCP	Hazard Analysis Critical Control Point: a check point system used by food processors and food handlers to ensure that potential contamination is prevented or eliminated during the preparation or production of foods.
HARBORAGE	1.) shelter and breeding areas for pests. 2.) a human harboring a contagious disease as a carrier.
HELMINTH	a parasitic intestinal worm.
HEPATITIS A	a pathogenic virus. It is a contagious infection that can be transmitted through food by food handlers with poor personal hygiene.

HERMETICALLY SEALED CONTAINER	a package that does not permit the transfer of contamination or gasses between food and its surroundings.
HIGH RISK PERSON	one who may be susceptible to the adverse effects of foodborne illness; i.e. pregnant, immuno-compromised, elderly or an infant.
HOST	a person, animal, insect or plant in which another organism lives.
HOT HOLDING	the temporary storage or display of hot, potentially hazardous foods at an internal temperature of a minimum of 60° C (140°F) or hotter after cooking or reheating.
IF IN DOUBT RULE	the principle of discarding food if there is any doubt as to its safety or wholesome ness, e.g. If In Doubt, Throw It Out!
IMMUNO-COMPROMISED	a person with a dibilated immune system due to an existing disease or condition. e.g. pregnancy.
INCUBATION PERIOD	the time it takes for symptoms to appear after being infected.
INFESTATION	the presence of pests.
INSECTICIDE	a poison used for killing insects or their eggs.

LISTERIA MONOCYTOGENES   a facultative pathogenic bacteria found in soil. It grows well at low temperatures and causes the foodborne infection Listeriosis. Especially dangerous to an immuno-compromised person or pregnant woman.

MEDIUM   an environment in which pathogenic microorganisms can grow.

MICROBE   life form that may only be seen with a compound microscope, e.g. bacteria, protozoa, molds, viruses, etc.

MICROORGANISM   microbe.

MODIFIED (MAP) ATMOSPHERE PACKAGING   food packaged in an atmosphere that is different from air (i.e. a vacuum or a gas).

MONOSODIUM GLUTAMATE   a food additive used in some foods as a flavor enhancer. Some persons may be sensitive to Monosodium Glutamate.

MOLD   a fungi. Some molds spoil food. Molds reproduce by forming microscopic spores. Some may form toxins in food that are pathogenic to humans. Others are beneficial and are used to produce certain cheeses.

MUTAGENIC   mutation causing.

MYCOTOXINS   toxins produced by molds. Food on which molds have grown can contain pathogenic mycotoxins. Foods that have been associated with mycotoxins include: fruit juice, yogurt, sour cream, cheese, bread, grains, cured meats and jams; also includes aflatoxin associated with peanuts and tree nuts.

ORGANISM   any living thing.

PARALYTIC SHELLFISH POISONING   caused by a group of toxins elaborated by planktonic algae upon which the shellfish feed. There are 20 toxins responsible for paralytic shellfish poisoning (PSP), all of them derivatives of saxitoxin.

PARASITE   a pathogenic organism living on or inside a host and is dependant on the host for nutrients, e.g. tape worms or round worms. They may also be microorganisms, e.g. bacteria, yeasts, protozoa or viruses.

PARTS PER MILLION   a measure of chemical solution concentrations. (PPM) e.g. 100 milligrams of household bleach diluted with one liter (1 million milliliters) of water is expressed as 100 parts per million of household bleach.

PASTEURIZATION   a process which exposes food for a specific combination of time and temperature to destroy pathogens in milk or other foods and slows down the growth of other organisms that may cause spoilage.

PATHOGEN	any disease causing microorganism or toxin.
PERISHABLE FOODS	foods that spoil or decay quickly if improperly stored.
pH	the symbol for the concentration of hydrogen ion. The acidity or alkalinity of a product or substance is measured on a scale of 1.0 to 14.0, with 0 being most acidic, 7.0 being neutral and 14.0 being most alkaline (caustic or base).
PHOTO-DEGRADATION	caused by phenolic compounds which increase sensitivity to light and are not destroyed by cooking. These compounds are highly mutagenic and carcinogenic.
POTABLE	water which is safe to drink.
POTENTIALLY HAZARDOUS FOOD	food that allows pathogenic microorganisms to grow or produce toxins if it is contaminated.
PROTOZOA	tiny, simple, one-celled animals that can also be parasitic; found in soil, fish, meat and surface water. Most protozoa are harmless while others cause infection in humans and animals.

PUFFERFISH POISONING
poisoning caused by consuming members of the Tetradon family of toxins is one of the most violent intoxications from marine species. The gonads, liver, intestines and the skin of pufferfish can contain levels of tetradotoxin sufficient to produce rapid and violent death. Other names include Tetradon poisoning or Fugu poisoning. The flesh of many pufferfish may not usually be dangerously toxic.

RANCIDITY
degradation of food quality due to enzyme activity resulting in the oxidization of fats in foods. Rancidity causes an unacceptable change in flavor of foods and the oxidative form, affecting vegetable and animal fats, is thought to be carcinogenic and mutagenic.

RECONSTITUTED
dehydrated food products combined with liquids such as milk or water. These foods may become potentially hazardous when reconstituted but are usually considered safe foods when properly dry stored.

REFRIGERATION
the storage of perishable, fresh or potentially hazardous foods at a maximum internal temperature of 40°F (4 °C) or less.

RETORT	a device used to preserve sterilized food in a hermetically sealed container. It involves the following three steps:

1.) sterilization of the food product by flash heating and cooling in a tubular type of heat exchange system.
2.) aseptic filling of the relatively cool, sterile product into the sterile container.
3.) application of a sterile cover to the filled container and sealing the container in an atmosphere of either saturated or superheated steam.

RODENTICIDE	a poison used for killing rats or mice.
SALMONELLA	a facultative pathogenic microorganism (bacteria) that does not form spores. It causes the foodborne infection salmonellosis. It is found in beef, poultry, eggs, pork, raw milk, fish, reptiles, pets, rodents and humans. Over 2,000 serotypes have been identified.
SANITATION	the art and science of maintaining conditions that enhance good health.
SANITIZE	procedures used to control the growth of pathogenic microorganisms on clean food contact surfaces.

SANITIZER
a chemical compound used to kill pathogenic microorganisms on clean food contact surfaces. Common sanitizers used in the food industry are:

1.) Chlorine (Household Bleach);
2.) Iodine (Iodophors); and
3.) Quaternary ammonium (Q.U.A.T.S.).

Hot water is also a sanitizer when used at specific temperatures for a specified length of time.

SCOMBROID POISONING
caused by the ingestion of high levels of histamine formed when producing products such as swiss cheese or by spoilage of fishery products, particularly mahi-mahi or tuna.

SHELF LIFE
the length of time a food product can be properly stored without compromising food safety or quality.

SHIGELLA
a facultative pathogenic microorganism (bacteria) that does not form spores. It is primarily found in humans. It causes the foodborne toxicoinfection shigellosis, also known as bacillary dysentery.

SOLANINE
one of several cholinesterase inhibitors which occur naturally in some foods (green potatoes). These inhibitors inactivate butyrylcholinesterase which may be a body's first natural line of defense against poisons that are eaten or inhaled.

SPORE   in bacteria an encasement of a cell's
                               genetic information and materials
                               necessary to exist in a dormant state under
                               extreme conditions and to produce a
                               living bacterial cell when conditions for
                               viable reproduction are met.

                               in the case of mold or yeast a similar
                               generalized statement may be made.

STAPHYLOCOCCUS   a facultative, pathogenic microorganism
AUREUS   (bacteria) found in humans that produces
                               a "toxin" that is not destroyed by heat.
                               This causes the staphylococcal foodborne
                               intoxication.

STERILIZE   procedures used to destroy all
                               microorganisms.

SULPHITING   food additives used to preserve freshness
AGENTS   and color in some foods, wines and
                               pharmaceuticals. They are a chemical
                               hazard when consumed by persons who
                               are allergic to them, specifically persons
                               with asthma.

TIME AND   Hot food must be kept at an internal
TEMPERATURE   temperature above 140 °F (60°). Cold
RULES   food must be kept at below 40 °F (4 °C)
                               and must not be allowed to remain in the
                               DANGER ZONE temperature range
                               between 40°F (4°C) 140°F (60 °C) for
                               more than an accumulated period of two
                               hours.

                               Frozen food must be kept at less than
                               0°F (-18°C) and thawed at temperatures
                               less than 40°F (4°C) or cooked while it is
                               still frozen.

TOXINS poisons, produced by pathogenic toxin forming microorganisms, e.g. *Staphylococcus, C. perfringens, Shigella, C. botulinum, E. Coli*, etc.

TRICHINELLA SPIRALIS a roundworm parasite found in hogs, bear, walrus and wild game that causes the foodborne illness Trichinosis.

VEGETATIVE CELL a microorganism which can grow and reproduce.

VENTILATION removal of heat, steam, grease and smoke from the food preparation areas and replacement with clean make up air.

VIABLE microorganisms or spores capable of living and being vegetative.

VIBRIO PARAHAEMOLYTICUS aerobic pathogenic bacteria may contaminate seafood and cause gastroenteritis if uncooked or lightly cooked seafood is consumed.

VIBRIO VULNIFICUS aerobic pathogenic bacteria found in water, sediment, plankton and shellfish, this bacteria infects only humans and other primates causing wound infections, gastroenteritis or a syndrome known as "primary septicaemia".

VIRUS the smallest of microorganisms. Viruses are pathogenic to humans and require the host's cells to replicate. They are known as intra-cellular parasites. They may be transmitted to humans through food, e.g. Infectious Hepatitis A foodborne infection.

YEAST

a fungi. Some yeast microorganisms are agents that cause food spoilage. Beneficial yeasts are used in the production of bread and beer. Yeasts reproduce microscopically by a process known as "budding". Some yeasts are pathogenic to humans.

YERSINIA ENTEROLITICA

facultative bacteria not normally a part of the human flora but often isolated from animals such as pigs, birds, beavers, cats and dogs. Most isolates have been found not to be pathogenic but it has been detected in water and food and may cause gastroenteritis.

ZOONOSIS

diseases normally found in animals which may be occasionally transmitted to man, such as Anthrax.

References and Recommended Reading

Animals Parasitic in Man, Geoffrey Lapage, Penguin Books, Ltd., Baltimore, U.S.A., 1957.

Bernardin Guide to Home Preserving, Bernardin Ltd., Toronto, Ontario, 1995.

Canned Foods, Ministry of Supply and Services Canada, 1992.

A Clinician's Dictionary of Bacteria and Fungi, James H. Jorgensen, PhD, and Michael G. Rinaldi, PhD, Eli Lilly and Company, Indianapolis, Indiana, 1986.

Control of Communicable Diseases Manual, Abram S. Benenson, American Public Health Association, Washington, DC, Sixteenth Edition, 1995.

Fish Safety Notes, B.C. Ministry of Health and Ministry Responsible for Seniors, 1993, 1994, 1995.

Food Additive Pocket Dictionary, published by the Minister of Supply and Services Canada, 1992.

Food Additives: Questions and Answers, Health Protection Branch, published by the Minister of Supply and Services Canada, 1990.

Food Code, U.S. Department of Health and Human Services, Public Health Service, Food and Drug Administration, Washington, DC, 1995.

Food Protection, Vital to your Business, B. C. Ministry of Health and Ministry Responsible for Seniors, 1989.

Food Sex & Salmonella, The Risks of Environmental Intimacy, by

David Waltner-Toews, DVM, NC Press Limited, 1992.
Food Science, by N.N. Potter and J.H. Hotchkiss, Chapman & Hall, New York, New York, Fifth Edition, 1995.

Foodborne Diseases, Edited by Dean O. Cliver, Academic Press, Inc, London, U.K., 1990.

Foodborne Pathogens: Risks and Consequences, Task Force Report No. 122, Council for Agricultural, Science and Technology, Ames, Iowa, September 1994.

Joy of Cooking, Irma S. Rombauer and Marion Rombauer Becker, New American Library, New York, New York, 1931- 1964.

Microbial Food Poisoning, Edited by Adrian R. Eley, Chapman & Hall, London, U. K., 1992.

Modern Food Microbiology, James M. Jay, Chapman & Hall, New York, New York, Fourth Edition, 1992.

Nutrition Labeling Handbook, prepared by Food Division, Consumer Products Branch, Consumers and Corporate Affairs Canada, 1992.

A Pocket Guide to Can Defects, The Association of Food and Drug Officials, York, PA, U.S.A.

Putting Food By, Ruth Hertzberg, Beatrice Vaughan, Janet Greene, Bantam Books Inc., New York, New York, 1973-1980.

Safe Food, Eating Wisely in a risky World, by Michael F. Jacobson, Ph.D., Lisa Y. Lefferts and Anne Witte Garland, Living Planet Press, Los Angeles, U.S.A., 1991.

The Technology of Food Preservation, Norman W. Desrosier, Ph.D., The Avi Publishing Company, Inc., Westport, Connecticut, Third Edition, 1970.

Resources and Consumer Information

Australia and New Zealand

Australia New Zealand Food Authority
P.O. Box 7186
Canberra Mail Centre
ACT 2610
Telephone 06-271-2222

Australian Quarantine and Inspection Service
Department of Primary Industries and Energy
G.P.O. Box 858
Canberra, ACT 2601
Telephone 06-272-5223

Australian Food Council
Locked Bag 1
Kingston, ACT 2604
Telephone 06-273-1466

National Farmers Federation
14-16 Brisbane Avenue
Barton Canberra, ACT 2600
Telephone 06-273-3855

Meat and Allied Trades Federation
P.O. Box 1208
Crowsnest NSW 2065
Telephone 02-9906-7767

Australian Dairy Products Federation
6th Floor, 84 William Street
Melbourne,
VIC 3000
Telephone 03-9642-8033

Canada

Health Canada
Health Protection Branch
Microbiology Research Division
Bureau of Microbial Hazards
4th floor west, Banting Bldg.
Tunney's Pasture
Ottawa, Ontario
K1A 0L2 Postal Locator 2204A2
Telephone 613-957-0880

Agriculture and Agri-Food Canada
Sir John Carling Building
930 Carling Avenue
Ottawa, Ontario K1A 0C5
Telephone 613-759-1000

Fisheries and Oceans
Centennial Towers
300 Kent Street, 14th floor
Ottawa, Ontario K1A 0E6
Telephone 613-993-0999

Each province and territory has an Enquiry Service number which can put you in touch with the appropriate department or trade organization.

Beef Information Centre
#100, 2233 Argentia Road
Mississauga, Ontario L5N 2X7
Telephone 905-821-4900

Canadian Egg Marketing Agency
Suite 1900, Place de Ville
320 Queen Street
Ottawa, Ontario K1R 5A3
Telephone 613-238-2514

Dairy Farmers of Canada
75 Albert Street, Suite 1101
Ottawa, Ontario K1P 5E7
Telephone 613-236-9997

Canada Pork Council
1101-75 Albert St
Ottawa, Ontario K1P 5E7
Telephone 613-236-9239

Canadian Turkey Marketing Association
969 Derry Road East Unit 102
Mississauga Ontario L5T 2J7
Telephone 905-564-3100

Canadian Chicken Marketing Agency
377 Dalhousie Street, Suite 300
Ottawa, Ontario K1N 9N8
Telephone 613-241-2800

Canadian Produce Marketing Association
1101 Prince of Wales Drive, Suite 310
Ottawa, Ontario K2C 3W7
Telephone 613-226-4187
Fresh fruit and vegetable information hotline
1-800-668-7763

United Kingdom

Ministry of Agriculture, Fisheries and Food

These help lines in MAFF will put you in touch with the right person to answer your questions.

Consumer Help Line
Room 306A, Ergon House
c/o Nobel House
17 Smith Square
London, SW1P 3JR
Telephone 0345-573-012

Deals with enquiries on food safety and consumer protection issues.

MAFF Help Line
Whitehall Place (West Block)
London, SW1A 2HH
Telephone 0645-335-577

Deals with enquiries on agriculture, farming and fishing.

United States

Food and Drug Administration
Center for Food Safety and Applied Nutrition
Division of Cooperative Programs
HFS-625
200 C Street, S.W.
Washington, DC 20204
Telephone 202-205-8140

Food and Drug Administration
5600 Fisher's Lane
Room 12-17
Rockville, Maryland 20857
Seafood Hotline 1-800-332-4010

United States Department of Agriculture
Food Safety and Inspection Service
Room 2925 South
1400 Independence Avenue, S.W.
Washington, DC 20250
Meat and Poultry Hotline
1-800-535-4555

The United States Department of Agriculture also provides extension services through universities in each state.

American Egg Board
1460 Renaissance Drive, Suite 301
Park Ridge, Illinois 60068
Telephone 847-296-7043

National Broiler Council
1155 15th Street N.W. Suite 614
Washington, DC 20005-2706
Telephone 202-296-2622

National Turkey Federation
1225 New York Avenue, N.W., Suite 400
Washington, DC 20005
Telephone 202-898-0100

National Pork Producers
P.O. Box 10383
Des Moines, Iowa 50306
Telephone 515-223-2600

Produce Marketing Association
P.O. Box 6036
1500 Casho Mill Road
Newark, Delaware 19714-6036
Telephone 302-738-7100

National Cattleman's Beef Association
Beef and Veal Culinary Center
444 North Michigan Avenue
Chicago, Illinois 60611
Telephone 312-467-5520

National Dairy Council
10255 West Higgins Road, Suite 100
Rosemount, Illinois
60018-5616
Telephone 708-803-2000

About the Editors

Wm (Bill) Hines, C.P.H.I.(c) R.E.H.O. Food Safety Consultant and Public Health Inspector, Community and Environmental Health Protection, Capital Regional District Health Department in Victoria, British Columbia.

Bill began his career as a Public Health Inspector in 1972 in Northern Saskatchewan. He is the founder of the FOODSAFE Sanitation Training Program For Foodhandlers, published in 1987. He was responsible for the conception and then the development and implementation of the FOODSAFE Program as a member of the FOODSAFE Steering Committee.

In 1992, he was appointed as Food Safety Consultant for the Capital Regional District. In 1994, he was responsible for food safety at the Commonwealth Games, held that year, in Victoria.

He was recently invited to act as the World Heath Organization Consultant to the Ministry of Health in Malaysia in developing a plan for food safety at the 1998 Commonwealth Games.

Bill continues to work at a grass-roots level as a Public Health Inspector and Food Safety Consultant in Victoria, British Columbia.

About the Editors

Gerry Penner, C.P.H.I.(c) Manager, Industry Training, Self-Monitoring and Liaison, Environmental Health and Safety Branch, Ministry of Health, British Columbia.

Gerry began his career as a Public Health Inspector with the British Columbia Ministry of Health in 1964. In 1981, he became a consultant for the Ministry of Health in environmental health programs with a special interest in food sanitation programs.

In 1985, Gerry was designated as the Ministry of Health representative on the Steering Committee established to develop the FOODSAFE Sanitation Training Program for foodhandlers, and he was responsible for the implementation and promotion of FOODSAFE education and certification programs.

He developed a national-award winning self-monitoring certification program for food services operators called "Achieving FOODSAFE Excellence" implemented in 1996.

Currently, Gerry is responsible for the development and implementation of safety education and self-monitoring programs for swimming pool operators in British Columbia based on the FOODSAFE program model.

About the Author

Sheri Nielson is an award winning instructor of the B.C. FOOD-SAFE Program. A former print and television journalist, she is also co-author of Running Your Restaurant: An Operations Manual, which was published in 1988 by the Restaurant and Foodservices Association of B.C.

Through her company, Quanta Restaurant Systems Ltd., she has certified over 8,000 people in FOODSAFE training since 1988. She is continuing her career in the food industry by working as a training consultant to restaurants.

She is also a licenced EMA 1 paramedic working part-time for the British Columbia Ambulance Service in her home community and an authorized provider of Canadian Red Cross first-aid courses. She is a former auxiliary police officer with the Ganges Detachment of the R.C.M.Police.

Her first questions about food safety surfaced as a young mother and homemaker. As with many people in the 1970's, she moved back to the "country" and discovered that she had only common sense and some old recipe books to fall back on. It was not until she became involved in the FOODSAFE program, that she started to find the answers to the questions that had been concerning her about food safety.

Sheri lives with her son on Salt Spring Island, midway between Vancouver and Victoria in British Columbia.